MW00592738

Fire Protection Systems

Second Edition

Justin Duncan

American Society of Plumbing Engineers
8614 W. Catalpa Ave., Suite 1007
Chicago, IL 60656

American Society of Plumbing Engineers
8614 W. Catalpa Ave., Suite 1007
Chicago, IL 60656
(773) 693–2773 • Fax: (773) 695–9007
E-mail: aspehq@aspe.org • Internet: www.aspe.org

Fire Protection Systems, 2nd edition, has been written as a general guide and is designed to provide accurate and authoritative information. Neither the author or publisher have liability nor can they be responsible to any person or entity for any misunderstanding, misuse, or misapplication that would cause loss or damage of any kind, including loss of rights, material, or personal injury, or alleged to be caused directly or indirectly by the information contained in this book.

The publisher makes no guarantees or warranties, expressed or implied, regarding the data and information contained in this publication. All data and information is provided with the understanding that the publisher is not engaged in rendering legal, consulting, engineering or other professional services. If legal, consulting or engineering advice or other expert assistance is required, the services of a competent professional should be engaged.

Copyright © 2001 by American Society of Plumbing Engineers

All rights reserved, including rights of reproduction and use in any form or by any means, including the making of copies by any photographic process, or by any electronic or mechanical device, printed or written or oral, or recording for sound or visual reproduction, or for use in any knowledge or retrieval system or device, unless permission in writing is obtained from the publisher.

ISBN:1–891255–14–2
Printed in the United States of America

10 9 8 7 6 5 4 3 2 1

Dedication

To the memory of my parents: Thank you for providing me with a solid life foundation.

Acknowledgments

Thanks to my devoted friend, L. Gurevici, P.E., for his continued support and active contribution.

Contents

Preface ix

Chapter 1:
What Is Fire Protection? 1

Chapter 2:
Basic Chemistry and Physics of a Fire 15

Chapter 3:
Fire Safety in Building Design 23

Chapter 4:
Fire Detection Systems 29

Chapter 5:
Fire Suppression 41

Chapter 6:
Fire Pumps 45

Chapter 7:
Water Suppression Systems (Manually Operated) 59

Chapter 8:
Automatic Sprinkler Systems 65

Chapter 9:
Hydraulic Principles and Properties of Water 89

Chapter 10:
Sprinkler System Calculations 113

Chapter 11:
General Information About Fire Protection Systems 135

Chapter 12:
Fire Protection System Inspection and Maintenance 145

Chapter 13:
Carbon Dioxide (CO_2) and Halon Replacement 151

Chapter 14:
Foam and Other Extinguishing Agents 165

Chapter 15:
Portable Fire Extinguishers 169

Chapter 16:
A Few Final Words on Fire Protection 173

Chapter 17:
Sample Specification for a Water Fire Protection System . . 175

Chapter 18:
Sample Specification for Portable Fire Extinguishers 213

Chapter 19:
Sample Specification for a High-Pressure Carbon
Dioxide Fire Protection System 217

Chapter 20:
Sample Specification for a Clean-Gas
Fire-Protection System 227

Appendix A:
Alphabetical Listing of NFPA Standards 239

Appendix B:
Characteristics of Water Flow in Pipes 247
Part 1: Friction of Water in Pipes 247
Part 2: Friction Losses in Pipe Fittings 261
Part 3: Corresponding Pressure Table 267
Part 4: Symbols 268

Appendix C:
Part 1: Pre-Installation Designer and Contractor's
Information List 275
Part 2: Sample of Fire Protection Notes 279

Appendix D:
Portable and Wheeled Fire Extinguisher Guide 281

Appendix E:
Sample of a Computer Hydraulic Calculation 289

Appendix F:
Sprinkler Room Arrangement, Backflow Preventer Detail . 295

Appendix G:
Extinguisher Selection Chart 299

Appendix H:
Unit and Conversion Factors 301

References 307

Index 309

Preface

Uncontrolled fires are dangerous. Every effort should be made to educate the public at large about the prevention, alarm, and suppression of uncontrolled fires.

Fire protection is a well-defined discipline that includes prevention, detection, and suppression. It differs substantially from other technical fields because fires can happen unexpectedly and be extremely dangerous. The National Fire Protection Association (NFPA) issues standards, which serve as a guide for fire prevention and cover any situation and industry segment. In spite of this continuous effort, though, fires still occur.

Fire Protection Systems, Second Edition, describes a wide variety of up-to-date firefighting possibilities and many available means of protecting against uncontrolled fires. It is my hope that this book will facilitate the understanding of fire protection as a whole by providing technical information in an easy-to-understand format.

—Justin Duncan

Chapter 1: What Is Fire Protection?

This book aims to provide the reader with an overall view of fire protection. It includes state-of-the-art information pertaining to fire-suppression systems, prevention, and control.

Uncontrolled fires are dangerous to people and property. Fire protection is a science based on facts, which evolves based on gained experience and improved equipment and tactics. Every person involved in fire protection (which includes prevention, suppression, detection, alarm, and system installation) should be familiar with fire-protection standards and applicable codes. The standards are issued by the National Fire Protection Association (NFPA), and the codes are issued by the governing state, with certain specifics supplied by counties and/or cities.

Authorities Having Jurisdiction

It must be clearly understood that the applicable code, or any governing code, does not abrogate, nullify, or abolish any law, ordinance, or rule adopted by the local governing authority having jurisdiction. According to the NFPA, the authority having jurisdiction is the organization, office, or individual responsible for approving an installation, equipment, or procedure.

Authorities related to fire protection may be federal, state, local, or other regional departments or an individual such as the fire commissioner, fire chief, fire marshal, or electrical inspector. An insurance inspection department, rating bureau, or other insurance company representative may also be an authority having jurisdiction. In many circumstances, the property owner, or a designated agent, assumes the authority role. At a government installation, it may be the commanding officer or departmental official.

The authority having jurisdiction may also refer to the listings or labeling practices of an organization concerned with product evaluations. One such organization is Underwriters' Laboratories (UL). UL publishes an annual list or directory of all the equipment it tests and approves. Factory Mutual (FM) publishes a similar document.

It should be noted that NFPA does not approve, inspect, or certify any installations, procedures, equipment, or materials, nor does it approve or evaluate testing laboratories. In determining the acceptability of installations, procedures, equipment, or materials, the authority having jurisdiction may base its acceptance on compliance with NFPA or other applicable standards and codes.

Before any fire-protection system is installed, extended, or remodeled, code dictates that a permit must be secured from the authority having jurisdiction. Approval for a construction project and the issuance of an occupancy permit is typically given by the building inspector. The permit is usually based on the fire marshal's approval of the fire-protection system(s), and the representative of the owner's insurance underwriter must also concur.

The "permit" is an official document issued by the authority having jurisdiction. It authorizes building a new construction or proceeding with the remodeling of an existing one and its fire-protection system. Permits are issued in the name of the owner for the use of storage, manufacturing, occupancy, or control of specific hazardous operations. The permit is issued if it is determined that the conditions will meet the minimum requirements of applicable code and standards.

The permit process provides the fire-prevention official with information as to what, where, how, and when a specific (hazard) system that is under his/her jurisdiction will be installed. Further, it allows the fire official to review and approve devices, safeguards, and procedures that may be needed to ensure the safe use of (hazardous) materials or operations. Permits are not transferable.

The first step in project development is presentation of the design concept to the authority having jurisdiction, who, in turn, issues the preliminary approval. When a fire-protection project is being developed, the following steps usually take place:

1. Initial project development (conceptual).
2. Project presentation for overall approval.
3. Construction/installation.
4. Inspection and tests.
5. Issuance of the occupancy permit.

Before the occupancy permit is issued, as well as during the project construction, periodic inspections may be performed by the building and/or fire inspector. Certain items usually need correction, such as the following:

- Elimination or supervised confinement of highly combustible materials.
- Elimination of combustible or hazardous materials that do not have special precautions in place.
- Violations of established fire-prevention rules.

After the construction (installation) is completed and the occupancy permit is issued, the owner (or representative) and the authority having jurisdiction must still be in contact. This is due to the fact that the owner must prepare and implement various plans for inspection. These include evacuation plans, fire-safety plans (implementations and improvements), and a schedule and record of fire drills.

These plans and schedules must be updated periodically, especially after changes in occupancy, physical rearrangement of the area, or technical developments. It is the owner's responsibility to promptly update the fire-safety and evacuation procedures in such instances.

General Precautions

The person supervising or installing the fire-suppression system and the person responsible for maintaining a building with such a system must be familiar with basic fire-protection requirements, which are necessary for life and property protection. Specific precautions are also part of these requirements and must be observed under the following conditions:

- During welding operations (permit required).
- To provide for safe storage of combustible materials (permit required).
- To prevent the accumulation of waste materials and promptly dispose of them.
- When an open flame is used for any reason (permit required).
- To establish measures required for reducing, replacing, or closely supervising any item connected with the above list of safety precautions

"Fire-protection systems" are engineered systems defined as functional units requiring individual calculations and design to determine the flow rates, nozzle pressure, quantities of extinguishers (water, chemicals, or gas), and the number and type of nozzles and their placement for protection and fighting a fire for a specific hazard.

The designer and the owner must know what is required and recommended for fire protection for a certain type of occupancy and a particular type of construction. This is based on practice and/or accumulated knowledge of the applicable standards and codes. However, NFPA lists the different types of occupancy and the types of hazard to which they belong (see Table 1-1). Table 1-2 summarizes the types of occupancy and their characteristics. The correct hazard designation is important because if the hazard is

Table 1-1   Occupancies and Hazards

Light-Hazard Occupancies

Churches	Museums
Clubs	Nursing or convalescent homes
Eaves and overhangs, if combustible construction with no combustibles beneath	Offices, including data processing
	Residential
Educational	Restaurant seating areas
Hospitals	Theatersand auditoriums, excluding stages and prosceniums
Institutional	Unused attics
Libraries, except large stack rooms	

(Table 1-1 continued)

Ordinary-Hazard Occupancies (Group 1)

Automobile parking and showrooms
Bakeries
Beverage manufacturing
Canneries
Dairy products manufacturing
  and processing

Electronic plants
Glass and glass products
  manufacturing
Laundries
Restaurant service areas

Ordinary-Hazard Occupancies (Group 2)

Cereal mills
Chemical plants—ordinary
Confectionery products
Distilleries
Dry cleaners
Feed mills
Horse stables
Leather goods manufacturing
Libraries—large stack room areas
Machine shops
Metal working
Mercantile

Paper and pulp mills
Paper-processing plants
Piers and wharves
Post offices
Printing and publishing
Repair garages
Stages
Textile manufacturing
Tire manufacturing
Tobacco-products manufacturing
Wood machining
Wood-product assembly

Extra-Hazard Occupancies (Group 1)

Aircraft hangars
Combustible hydraulic
  fluid use areas
Die casting
Metal extruding
Plywood and particle board
  manufacturing
Printing (using inks having flash
  points below 100°F (37.9°C)

Rubber reclaiming, compounding,
  drying, milling, vulcanizing
Saw mills
Textile picking, opening, blending,
  garnetting, carding, combining of
  cotton, synthetics, wool shoddy, or
  burlap
Upholstering with plastic foams

Extra-Hazard Occupancies (Group 2)

Asphalt saturating
Flammable liquids spraying
Flow coating
Mobile home or modular building
  assemblies (where finished
  enclosure Is present and has
  combustible interiors)

Open oil quenching
Plastics processing
Solvent cleaning
Varnish and paint dipping

Table 1-2   Occupancies and Their Characteristics

Occupancies	Quantity of Contents	Combustibility of Contents	Heat-Release Rate	Stock-Pile Height	Use of Flammable or Combustible Liquids
Light hazard	Low	Low	Moderate	None	None
Ordinary Hazard (Group 1)	Moderate	Low	Moderate	Less than 8 ft	Very limited
Ordinary Hazard (Group 2)	Moderate to high	Moderate to high	Moderate to high	Less than 12 ft	Very limited
Extra Hazard (Group 1)	Very high	Very high	High*	See NFPA standards	Moderate amount
Extra Hazard (Group 2)	Very high	Very high	High*	See NFPA standards	Large amount

*Also danger of rapid fire development.

underestimated, the suppression system might not be adequate for the particular combustible content.

The fire-protection discipline is generally divided into two distinct parts: fire prevention and fire suppression. However, a third element, fire detection and alarms (usually handled by the electrical engineer), is also part of fire protection. Included under fire prevention is the behavior and readiness of individuals and their reaction to prevent the loss of life in case of a fire. To maintain good response capability, all building occupants must be kept up-to-date regarding fire-safety procedures.

Fire-Safety Education

Continuous education is the very heart of fire prevention. Without education, prevention is not possible. The person responsible for fire protection must convey to the people who work in or occupy a building that they need to work together toward fire prevention and protection. Fire safety is part of real life education.

In 1987, according to NFPA, public fire departments responded to a total of 2,330,000 fires, or approximately 6400 per day. In the same year, the total number of civilian deaths in the United States attributed to fire was 5800, and total property damage was more than $7 billion. Remember that not all fires were reported or included in this statistical sample.

Research shows that when people receive training or instructions on what to do in case of fire, they are more likely to sound the alarm and organize an evacuation. It was determined that the psychological and behavioral effects of fire upon human beings impede a safe escape because among these effects is physical incapacitation due to panic, resulting in faulty judgment. Researchers postulate that the behavior of people in the case of a fire can be understood as a logical attempt to deal with a complex, rapidly changing situation in which minimal information is available.

Children also must be methodically taught that uncontrolled fires are dangerous. It is our responsibility as adults to educate children continuously how to be safe in case of a fire.

In the first decade of the twentieth century, no technical committee was exclusively geared toward life-safety concerns. The Triangle Shirtwaist fire of March 25, 1911, changed that, and helped in the development of today's *Life Safety Code*.

One of the largest clothing manufacturing companies in New York City, the Triangle Shirtwaist Company was located on the 8th, 9th, and 10th floors of the Asch Building. There were more than 500 employees in the company, many of whom were young women and immigrants, who worked long hours in dirty, cramped conditions.

The building itself was a firetrap. It was constructed nearly completely of wood, which was unusual for a building as tall as it was. Instead of three stairways, as were required by city codes, the building had only two. The architect had pleaded that the fire escape outside the building could suffice as the third stairway. The fire escape, however, went only as far as the second floor. The doors to the exits opened in toward the rooms instead of outward because the stairway's landing was only a stair's width from the door. Also, egress routes were narrow and full of obstacles, and partitions were placed in front of elevators and doors. Finally, the Triangle's housekeeping contributed to the fire. Rags from cutaway cloth materials frequently piled up on the floor and in storage bins. At the time of the fire, the rag bins had not been imptied in two months.

Just before quitting time on March 25, 1911, a worker noticed smoke coming from one of the rag bins. In the clothing industry, a fire of this nature was not unusual, but this fire spread rapidly, overcoming employees who tried to put out the fire with buckets of water.

Workers on the 8th floor rushed for the exits. One exit was locked, a company policy during working hours. Once it was unlocked, a panic ensued, causing a logjam of people in the stairway. Other workers frantically ran for the elevators, but the elevators had been summoned to the 10th floor, where the executive offices were located. When the elevators arrived, they were crammed with people. The elevators made so many trips in an effort to save workers on the 8th and 10th floors that the operators were finally overcome by smoke and exhaustion. Some workers climbed out onto the fire escape. One person fell down the fire escape to the courtyard below. Others climbed down to the 6th floor then down the stairs to the street.

There were approximately 260 workers on the 9th floor, which was congested with long sewing tables that ran along the length of the floor. The only way to exit was to walk all the way to one end, negotiating around chairs and baskets. When the quitting bell rang, the first worker out walked down the stairs to go home. When he reached the 8th floor, he noticed smoke and flames. He continued on a short distance then realized that he must warn the others on the 9th floor. By then, however, it was too late. The stairs leading back to the 9th floor were consumed in flames.

The 9th-floor workers discovered the fire when it entered the windows from the floor below. About 150 workers raced for the remaining stairway and about 100 made it to the street. Others ran for the fire escape. Jammed with people and hot from the fire, the fire escape pulled away from the building, sending many people to their deaths. Many others rushed for the elevators, but they were full. Some jumped or were pushed into the elevator shaft. A few slid down the elevator cables.

The fire department arrived in a timely manner, but could do little because its equipment only reached to the 7th floor. A total of 147 people lost their lives in the fire.

Fire Drills

It may appear childish to rehearse fire drills with adults in offices, institutions, and factories. To some it may seem unnecessary to instruct adults about safety, emergency exits, behavior in case of fire, panic, etc. However, the opposite is true. Fire drills are extremely important in the prevention of losses and should be repeated frequently in accordance with a safety plan. The following fire-drill requirements are reprinted from the Massachusetts fire-prevention regulations:

All hospitals shall conform to the following fire-drill regulations:

1. Each hospital shall formulate a plan for the protection and evacuation of all persons in event of fire: such plan shall be presented to and approved by the fire-department head. All employees shall be kept informed of their duties under such a plan.

2. The local fire-department head shall visit each hospital in his/her jurisdiction at least four times per year for the purpose of ascertaining whether the supervisors, attendants and other personnel are familiar with the approved evacuation plan.

A very similar set of regulations is valid for schools, offices, etc. The people involved in fire protection must participate when such material is prepared.

Airlines have specific safety instructions that are demonstrated to passengers every time the airplane leaves the ground. It is a strictly enforced regulation. Fire-protection drills should also be repeated regularly for building occupants, instructing them what to do (and not do) in case of a fire.

Another important aspect of fire protection is having a free and unobstructed escape route. Such routes have proved time and again to be life savers.

Fire-Protection Organizations

There are two main organizations (among many others) dedicated to saving lives and property from fire. They are the National Fire Protection Association (NFPA) and Factory Mutual (FM).

The NFPA

The NFPA[1] is a nonprofit, technical and educational organization dedicated to the protection of lives and property from fire. Founded in November 1896, the association is recognized today as the leading international advocate for fire safety. The association was founded when the need for a single standard regarding sprinkler installation in buildings was recognized. The association administers a standard-developing system and publishes fire-safety standards (see Appendix A for a list of NFPA standards[2]). These are used as a guide by fire-protection professionals, insurance companies, business, and government alike. The NFPA also provides fire information and statistics to the fire-protection field, conducts on-site investigations of significant fires, and develops publications, films, and training programs. These constitute the basis of education for the fire-protection community and the general public. The NFPA is a membership organization consisting of people in the fire service, design engineers, contractors, insurers, business and industry representatives, government officials, architects, educators, volunteers, and private citizens.

Initially, NFPA standards were directed and developed to consider only property protection. It did not take long to realize the need to include the protection of people, which is now the priority. **NFPA standards do not have the power of enforcement, they are strictly advisory.** However, these standards have been adopted as the basis for most of the applicable fire-protection codes, which have enforcing power.

The *National Electrical Code*, which governs the electrical field, is one of the NFPA's standards (NFPA Standard no. 70) and a good example of an NFPA standard guiding regulations for an entire field.

Over time, and with the NFPA's help, a group of people known as "fire-protection specialists" has developed. These specialists are involved in the fire-protection business and have learned to deal with every facet of fire protection. The NFPA staff, support personnel, and fire specialists (at large) examine how fires begin, provide guidance as to how they should be prevented, and demonstrate how the public can be better educated to reduce or avoid the very dangerous threat of uncontrolled fires.

Factory Mutual

Factory Mutual (FM) performs research and testing, offers guidance, and provides insurance in the fire-protection field. It was founded in 1835 when it became apparent that large industrial and commercial companies needed fire-insurance coverage. At the present time, the Factory Mutual system consists of three insurance firms—Allendale Insurance, Arkwright Insurance Company, and Protection Mutual Insurance Company.

The Factory Mutual Research Corporation conducts research and develops new and improved equipment, devices, and systems for the fire-protection industry. Factory Mutual is similar to Underwriters' Laboratories in that they test equipment, devices, and systems to determine if their reliability and efficiency will receive the "FM Approval" label. The *Factory Mutual Approval Guide* lists all the products, devices, equipment, and systems approved by FM. The guide also includes details of installation and materials.

Factory Mutual publishes its own requirements in *Factory Mutual Data Sheets*; however, some NFPA standards are adopted in their entirety. All the above descriptions of FM capabilities constitute an important part of fire-protection education and practical application.

Codes and Standards

Engineers and technicians design dependable fire-suppression systems, and contractors install them. Codes and standards give these professionals the guidance they need. The fire-protection specialist should have knowledge of the applicable standards and codes, know where to find a reference when required, enforce the codes, obey the laws, and be dedicated to improving them.

There are two major codes or standards applicable to fire protection: (1) the state building code, which regulates the type of construction (and all its derivatives), and (2) the state fire-prevention codes and standards. Building Officials and Code Administration International (BOCA) also issues fire codes, which incorporate part of the NFPA standards.

A "fire code" is defined as a set of rules (principles) and regulations adopted by the authority having jurisdiction to ensure minimum safety requirements against fires. A "standard" is defined as a set of recommended principles and guidelines established by a professional organization (e.g., NFPA) as the basis for the design, installation, and maintenance of a certain system. Fire-protection codes and standards were developed in order to protect the lives of occupants as well as the property and its contents.

Generally, the purpose of fire codes is to set minimum levels of acceptability in the design, installation, and maintenance of fire-protection systems. Most codes, as well as insurance-company requirements, establish performance objectives by providing specific requirements. These codes leave it up to the designer to determine how to meet those objectives. More than one solution is usually applicable, because new and original ideas are constantly being developed.

Fire codes do not permit building inspectors to grant waivers from code requirements. However, a safe alternate substitution may be acceptable and approval may be granted for such installation.

All local regulations applicable to fire protection and requested by the authority having jurisdiction are mandatory and/or enforceable. While prevention, protection, and firefighting are expensive, they are far less expensive than the destruction of unprotected life and property.

In Europe, a coordination program called the European Committee for Standardization is helping to develop general, common standards as well as fire-protection standards. These standards are based on French, German, and British national standards, and some 90 other countries are included.

Notes

[1]The world headquarters of the NFPA are located at 1 Batterymarch Park in Quincy, Massachusetts 02269, (617) 770-3000 or Customer Services 1-800-344-3555. In the United States, the NFPA has regional offices in Washington, DC; California; and Florida.

[2]This list is continuously updated with additional and revised standards. When needed, consult the latest edition available from the NFPA.

Chapter 2:
Basic Chemistry
and Physics
of a Fire

A fire is a chemical reaction involving fuel, oxygen, and heat. These elements form what is called the "fire triangle" (see Chapter 5). Chemical reactions can be either "endothermic," a reaction that consumes heat during the process, or "exothermic," a reaction that releases heat during the process.

Heat is the energy that is absorbed or emitted when a given chemical reaction occurs. In the case of fire, energy in the form of heat is required to begin the reaction, and then after the reaction is started, heat is released. In other words, combustion begins as an endothermic reaction and then continues as an exothermic reaction. In the case of an explosion, the combustion reaction proceeds rapidly.

Most combustibles, such as solid organic materials, flammable liquids, and gases, contain a large percentage of carbon (C) and hydrogen (H). The most common oxidizing material is the oxygen (O_2) found in air. Air is composed of oxygen (approximately 20%), nitrogen (approximately 80%), and traces of other elements. In general, any material containing carbon and hydrogen can be combined with oxygen, or oxidized.

Usually, both fuel and oxygen molecules must be brought together and then activated before a fire is produced. This activation can be caused by:

- A spark from a nearby fire or from electrical equipment.
- High friction between two hard surfaces rubbing together, which in turn, elevates the material's temperature.

- Intense heat, which creates the possibility of the material reaching the flash point (see definition in NFPA Standard no. 30, *Flammable and Combustible Liquids*).

Once the fuel and oxygen are combined and activated, a chemical chain reaction starts, which causes fire to develop. Heat, smoke, and gases are continuously produced during this process. Once the fire begins, it will continue to burn as long as there is fuel, oxygen, and heat present.

Other elements that may affect a fire include:

- A catalyst—the addition or subtraction of a substance that may affect the rate of chemical reaction, while the substance itself is not changed.
- Inhibitors or stabilizers—substances that hinder the mixing of fuel and oxygen.
- Contaminants—substances that, if present, may or may not influence the reaction.

Smoke

Combustion produces smoke, gases, and heat, which form what is called the "fire signature." The fire signature is never the same for two fires. Smoke, gases, and heat can produce drastic changes in the environment and be hazardous to humans.

Both people and property are heavily affected by smoke. NFPA Standard no. 92A, *Recommended Practice for Smoke Control Systems*, is a very good source of information on smoke. Statistics show that when a fire occurs about 60% of human casualties are due to smoke and toxic-gas inhalation. This may be due to confusion, since people reaching a smoke-filled area on the way to an escape route will normally turn back rather than go through the area to safety.

The American Society for Testing and Materials (ASTM) defines "smoke" as a complex mixture of "airborne liquid, solid particulates and gas evolved from the chemical reaction of combustion." These airborne particulates are lightweight, and they rise and spread by air movement.

The amount of smoke produced when a fire burns depends on the mass of air or gas drawn into the fire, which, in turn, is based on the type of combustible. The amount of air is based on the pressure difference between the fire area and the adjacent space.

Smoke does have a few useful purposes, such as preserving meats, producing colored military signals, and protecting orchards during freezes. However, smoke inhalation is detrimental to our health and can be lethal.

Smoke Control

Since the early 1970s, it has become evident that, in the design of multistory buildings, smoke control should be included as part of the life safety systems. In all buildings, buoyancy and the stack effect cause smoke to travel upward; however, smoke movement differs between short buildings and tall buildings. In a short building, the influences of heat convective movement and gas pressure are major factors in smoke movement. In tall buildings, the stack effect drastically modifies the same factors because there is a strong draft from the ground floor to the roof due to the difference in temperature.

Computerized smoke-control models have been developed to assess and/or control smoke movement in a building. These models can simulate the expected behavior of smoke in a multilevel building. Variables such as outside air temperatures, wind speed, building height, air leakage (in and out), building configuration, stack effect, thermal expansion, air supply, and air exhaust can all be programmed into a computer-simulated scenario. This modeling is useful in planning and assessing building design and performance.

The present trend of smoke control in buildings is to create smoke-free areas, such as a building's egress or stairwells. The method of stairwell pressurization is an accepted way to prevent smoke from seeping into the stairwell enclosures. However, care must be taken not to create too much over-pressure, which can make access into.the stairwell through doors nearly impossible. For this reason, doors are designed to open out of rather than into a stairwell. The stack-effect and air-movement influences are also factors in creating a smoke-free stairwell. Ducting air into the stairwell at different levels is desirable to prevent uneven pressurization.

The need for smoke control can be demonstrated by a fire that occurred on the 14th floor of Boston's 52-story Prudential Tower in January 1986. This building was built in the early 1960s when fire and smoke regulations were greatly lacking compared to today's standards. Sprinklers were not installed, and the building depended on duct-mounted smoke detectors to shut down the supply and return air of the affected area(s). The fire dampers installed in the ducts supplying conditioned air were normally held open by a fusible link that melted at a predetermined temperature and allowed the damper to close and isolate the affected area. Physical barriers, which included heavy steel and concrete walls and/or floors, were able to contain the fire's heat but not the smoke. Driven by air expansion, the smoke penetrated the building's perimeter ductwork and elevator shafts, and within minutes, smoke spread throughout the entire building. There were two stairwells for egress, one of which was designed as a relatively smoke-free area. The other stairwell depended on the natural stack effect to rid the building of smoke. In the confusion, people used both stairwells for egress, and propped doors open with chairs, which allowed smoke into the designated escape stairwell.[1]

Another concept for smoke control involves the pressurization capability of the floors above and below the space where fire occurs. This air-pressurized barrier prevents smoke from infiltrating the adjacent floors by producing a higher pressure than the floor where the fire and smoke developed. Such an arrangement can be programmed into the air-conditioning system as a "fire-emergency mode."

Chemistry

Fire-protection specialists must have some knowledge of chemistry. Knowing it helps the specialist estimate the combustibility of the materials in an area as well as the heat and smoke expected to develop during a fire.

The "combustibility" of a material really means its capacity to burn. Combustible materials often present themselves in the form of gases, liquids, and solids. Simple organic materials include common fuels, which are also the building blocks of more complex fuels. For example, organic liquids like solvents and hydraulic fluids are all highly combustible. Common combustibles encountered in everyday activity include the following:

- Wood and all wood products.
- Textiles and all textile materials.
- Cushioning, man-made foam, and other applicable synthetic materials.
- Finishes like paints, stains, lacquers, etc.
- Flammable liquids and gases.
- Plastic materials.

A "noncombustible material," as defined by NFPA, is "a material which, in the form in which it is used and under the conditions anticipated, will not ignite, burn, support combustion, or release flammable vapors, when subjected to fire or heat."[2] A building construction material of limited combustibility may be acceptable, as stated by the NFPA:

A building construction material, not complying with the definition of non-combustible material, which, in the form in which it is used, has a potential heat value not exceeding 3500 Btu per lb and complies with one of the following paragraphs (a) or (b). Materials subject to increase in combustibility or flame spread rating beyond the limits herein established through the effects of age, moisture, or other atmospheric condition shall be considered combustible.

(a) Materials having a structural base of non-combustible material, with surfacing not exceeding a thickness of ⅛ inch which has a flame spread rating not greater than 50.

(b) Materials, in the form and thickness used, other than as described in (a), having neither a flame spread rating greater than 25 nor evidence of continued progressive combustion and of such composition that surfaces that would be exposed by cutting through the material on any plane would have neither a flame spread rating greater than 25 nor evidence of continued progressive combustion.[3]

Organic solids, which are another category of combustible materials, are not part of construction materials.

As previously stated, the principal constituents of combustible materials are carbon (C) and hydrogen (H^+). Combustible organic solids are classified as either hydrocarbons, with the chemical compounds CH and CH_2 as a base, or others like cellulose and its compounds, which contain the chemical group CH (OH). When these materials burn, the resulting products are carbon dioxide (CO_2) and water (H_2O). If any of these combustible organic materials are present when a fire occurs, the flames propagate quickly (at a rate of a few feet per second).

Fire Extinguishing

In attempting to control a fire, the aim is to break the chemical reaction or the continuous combination of fuel and oxygen. Another goal is to reduce one of its products—heat. Since fire is an exothermic reaction, one way to extinguish a fire is by cooling.

The oldest and most universally known fire-extinguishing agent is water. Water works as an extinguishing agent because it:

- Absorbs heat: 1 gallon per minute (gpm) at 60°F can absorb 1000 Btuh.
- Can extinguish a fire in a closed area at a rate of 1 gpm to a volume of 100 ft^3.
- Vaporizes at 500°F and expands 2500:1 at this temperature.
- Is more effective when mixed with thinning agents, becoming what is referred to as "wet water."
- Reduces the heat generated by a fire.

Other ways to extinguish a fire or control the chemical reaction are as follows:

- Remove the fuel.

- Reduce or eliminate the oxygen available for combustion by introducing an inert gas such as nitrogen (N) or, in small fires, cover the fire with a blanket.
- Apply chemical extinguishers such as CO_2, sodium, or potassium bicarbonate (or other dry chemicals), halon replacement (or another of the new substitutes).

To prevent the occurrence and/or the spread of fire, the fire specialist should come up with different methods of reducing the combustibility of various materials. These methods may include (for unoccupied areas) creating an inert atmosphere or using fire-retardant materials. However, there are many materials that contain oxidizing agents. These agents will provide oxygen for combustion even in an inert atmosphere, so be aware of their presence.

Fire-retardant or flame-resistant treatment of otherwise combustible materials helps protect against fires. This type of treatment for textile or wooden materials substitutes or impregnates the material with a noncombustible (less combustible) substance. The process can be accomplished through either an absorption or a saturation process. Impregnation can be done in a vacuum, in which case it is called "pressure impregnation."

Another way to protect materials or lower their combustibility is by applying a fire-resistant coating. Some fire protection is better than none at all.

Notes

[1]Evidently there were not enough fire drills performed in this building, or people did not actively participate. Consequently, people did not respond as expected for life protection. Due to panic, the response was contrary to instructions.

[2]NFPA no. 220, *Types of Building Construction.*

[3]Ibid.

Chapter 3:
Fire Safety in
Building Design

Code requirements are always minimal; better protection, more precautions, and the inclusion of advanced techniques always cost more but produce better living conditions and a safer environment.

Fire safety must be incorporated early in the design stage of a building,[1] and the design must comply with the applicable building regulatory codes. One important element is the fire resistance of a building, which is detailed in NFPA Standard no. 220. The definition given for a fire-resistance rating is "the time, in minutes or hours, that materials or assemblies have withstood a fire exposure as established in accordance with the test procedures of NFPA Standard No. 251, *Standard Methods of Fire Tests of Building Construction and Materials.*" The minimum hourly fire rating is "that degree of fire resistance deemed necessary by the authority having jurisdiction."

All architectural and engineering disciplines involved in the design of a building project are also involved in various aspects of fire protection. This involvement includes:

- Location, number, and construction of normal and emergency exits, as determined by the architect.
- Design of emergency lighting (operated by battery or the employment of an emergency generator), fire-alarm systems and grounding, and the use of spark-proof equipment in hazardous locations, etc., as determined by the electrical engineer.
- Operation mode of the air-conditioning and/or ventilation equipment in case of fire, smoke-evacuation system, and design of fire-suppression systems, as determined by the mechanical engineer.

- Protection of the building support beams and columns against high heat, and competent structural calculations for the type of construction and protective material selections, as determined by the structural engineer.

Exits and Openings

During the design stage of a building, special attention is given to the protection of exits. This includes stairways, corridors, and exit doors. All stairs and other exits in a building should be arranged to clearly point in the direction of egress toward the street. Exit stairs that continue beyond the floor of discharge to the street should be interrupted at the floor of discharge by partitions, doors, or other effective means. For building construction, the applicable building code and NFPA standards should be consulted and the requirements strictly followed.

Building openings and penetrations are usually designed to help stop the spread of fire and smoke, while containing gaseous (CO_2 or a specific "clean agent" like Inergen), fire-extinguishing, flooding systems. If this is the type of extinguishment designed, then strategically located relief vents must be provided for the air displaced by the fire-suppression agents when they are released.

Fire Barriers

To contain a fire in a certain area, a building includes fire passive restraints or fire barriers. These are fire walls, fire-resistant floors, fire-rated doors, etc. Areas that may be more prone to fire, such as control rooms, cable-spreading areas, relay rooms, switchgears, computer rooms, and repair-and-maintenance shops, must be constructed of noncombustible materials. The walls, floors, and ceilings in these areas must also be designed with a fire rating per code requirement. For example, if a door must contain a glass opening larger than 100 in.^{2}, a different class of fire-door rating applies.

From a fire and smoke-protection point of view, doors are designed and constructed based on the degree of protection they provide, for example:

- Nonfire-rated doors, such as the type used in a one or two-family dwelling, which provide limited protection when closed.
- Fire-rated doors tested to withstand fire for a defined period of time (1, 2, or 3 hours).
- Smoke-stop doors made of lighter construction, which provide a barrier to the spread of smoke.

For industrial construction, automatic fire doors in walls must be used to cut off the following areas:
- Boiler room.
- Emergency or standby diesel-generator room.
- Oil-storage rooms.
- Storage rooms for combustible materials.
- Flammable, oil-filled circuit breakers, switches, or transformers within station, i.e., vaults.
- Fuel-oil pump and heater rooms.
- Diesel fire-pump room.

Fire-Safety Personnel

The prevention of fires involves a personnel network dedicated to enforcing codes and continuously educating the general public and themselves. Engineers, technicians, contractors, and firefighting personnel are the professionals who design, install, maintain, and operate fire-protection and fire-suppression equipment and systems. Everything should be ready for operation at any time for an unexpected fire.

Fire safety does not consist of fire prevention only. Because fires ignite continuously, it is necessary to learn how to fight or suppress a fire once it has started.

Every industry has its own specific fire hazards and its own danger points, but specially trained personnel help apply the right protection for the specific hazard. NFPA standards cover most industries and establishments and provide useful criteria for fire protection.

As a practical matter, every building (public or private) and business should include fire-suppression systems and develop a fire-prevention program to fit its specific needs. Occupants should become familiar with and practice the proposed life-saving features.

In large organizations and/or corporations, there is usually a person responsible for safety, which includes fire prevention. Such organizations should have a fire-loss prevention-and-control manager. This person should be dedicated to personnel safety and fire prevention by continuously upgrading fire-suppression capability.

New Construction

In the preliminary stages of building construction, there is a greater danger of fire. During this time permanent suppression means are not in place, but there are certain basic fire protection recommendations that should be followed:

- Provide a temporary water-supply[2] source (excluding salt, tidal, or brackish water) for fire protection during the initial construction period in the amount and pressure and residual pressure[3] required by the authority having jurisdiction. As construction progresses, the permanent water supply must be made available as soon as possible, and all temporary fire-protection water connections should be divorced from the permanent one. The main water supply can be fed from the yard fire-protection loop (if one is installed) or the city water connection and yard hydrants.
- Underground mains should be made available as soon as is practical, and temporary sprinklers should be installed and used until the permanent system is installed and charged.
- Standpipes should be brought up and maintained as construction progresses to be ready for firefighting use. For high-rise buildings, firefighting personnel prefer to have a standpipe (wet or dry) ready for operation, if needed, two floors below the highest floor that is ready.
- The use of open flames and welding/cutting equipment should be properly supervised. The observation or supervision of such operations should be continued half an hour after the work is completed. For such operations, temporary permits are usually required from the fire department.

- Weather shelters and dust covers should be flame resistant.
- Facilities for hydrant operation should be made available as soon as is possible, and emergency protection in the form of portable extinguishers and hose streams must be provided. In certain cases, a watchperson and standby firefighting apparatus are recommended.
- Combustible materials should be kept at a minimum. Form work, shoring, bracing, scaffolding, etc. should be made of mostly noncombustible materials, and the construction site should be kept clean and orderly. Contractor's sheds should be constructed of limited combustible materials or kept outside the new construction confines.
- On rock sites (when blazing for fire-protection lines), installation should be performed simultaneously with general excavation to avoid damage to newly placed concrete.
- Portable fire extinguishers should be made available within 100 ft of any work area and within 30 ft of welding, burning, or other heat-producing equipment.

In sum, when new construction is concerned, it is always smart to:

- Assign the overall fire prevention/protection to a responsible person.
- Expedite the installation of firefighting systems.
- Dispose of rubbish promptly.
- Store combustibles in enclosed, ventilated, easy-to-supervise areas.
- Closely supervise temporary heaters.
- Provide temporary fire-suppression equipment (e.g., mobile hose stations and portable extinguishers).
- Carefully handle flammable liquids and gases.
- Establish enclosed, controlled areas for smoking.
- Take special precautions during welding and other operations involving open flame.

Remodeling

During building alteration or remodeling, the sprinkler system should be reconnected or installed at an early stage and kept operational. If work is done on a certain section of the system, that section should be isolated while the rest of the fire-suppression system is kept operational. If the entire system is out of order, then standby fire apparatus and/or a watchperson may be employed per recommendations from the fire department or the authority having jurisdiction. After the system is repaired, refurbished, or modified, it must be reinspected and retested before the installation is considered complete.

In case a sprinkler system is rearranged (with no occupancy change) and sprinkler heads must be replaced, they should be in concurrence with the existing sprinkler's style, orifice diameter, temperature rating, coating (if any), and deflector type. All of these replacement criteria are true except if the occupancy and/or the type of inside construction details (e.g., ceiling removed or added) change.

Notes

[1]See NFPA Standards nos. 101 and 101M.

[2]A temporary water supply may be obtained from a river, lake, or other natural or man-made body of *fresh* water. In case a natural body of water is the supply source, the intake should be constructed to include debris-retention screens. (A means to prevent screen clogging with marine life should also be included and located below the lowest water level.) Pumps should be secured above the flood level, and these may feed directly into the permanent fire lines when those are installed.

[3]Residual pressure is the pressure remaining in a system after a drop in pressure due to the opening of a system's discharge valve.

Chapter 4:
Fire-Detection
Systems

As previously mentioned, the fire-protection system consists of prevention, suppression, detection, annunciation, and communication systems. This chapter concerns itself with the latter components, which include the following:

- Detectors sensing the products of a fire.
- Fire alarms and annunciators.
- Communication systems that activate fire-suppression systems or isolation devices.

Detection devices do not control or extinguish a fire; they merely detect fire products. Technological developments have resulted in a large variety of detection devices. A correct detection system must be properly designed and the detectors carefully selected for the types of fire and the resulting products expected, which depend on the combustible materials and operational activities within the area.

Even though detectors do not directly affect a fire, they may be connected to initiate other functions, including:

- Sounding a local and/or remote alarm, which notifies the building occupants of a fire situation.
- Isolating an area by closing dampers and doors.
- Either stopping ventilation equipment or starting smoke-evacuation fans and opening fresh-air dampers or doors.
- Supervising the system for "ready-for-operation" status.
- Activating fire-suppression systems.

Detectors in high-rise buildings or industrial complexes may also be connected to a local, central annunciator panel. Panels are usually located in a control room, which is usually continuously attended. The control panel may also receive "trouble signals," which indicate such things as a fault in the supervisory system, a component being in the wrong position, and sprinklers operating inadvertently.

Manual and Automatic Detection Systems

A detection system can be either manual or automatic. A manual system relies on a person to observe fire and/or smoke and pull an alarm to alert the occupants. The person may also activate a suppression system. An automatic system relies on a detector to detect products of combustion and activate an alarm or fire-suppression system and other auxiliary systems (smoke evacuation, etc.).

An automatic detection system notifies the building occupants of a fire condition and summons an organized response. It may also activate a fire-suppression system, supervise the protection system, and detect any signs of change.

Before installing an automatic detection system, it is first necessary to establish whether or not it is necessary. Local codes or regulations may provide guidance for this decision. Factors that affect the decision include:

- Importance of the area (types of content and their value).
- Susceptibility to a surrounding area fire hazard.
- Degree of fire hazard within the area.
- Potential of fire spreading.
- Type of fire suppression (e.g., chemicals, water).
- Normal occupancy of the area.
- Cost of detection system.
- Will system consist of detection and suppression or just detection?

Once a decision is made to install an automatic detection system, it is necessary to establish the detection requirements for the area. Then, select the appropriate detector type(s) and place them in the correct location and at the correct distance from one another.

Types of Detection Device

There are three basic types of detector: heat detector, smoke/gas detector, and flame detector. These have been proven to give a good practical response and are effective when installed in accordance with their governing qualification tests.

Heat Detectors

Heat detectors sense the heat produced by burning combustibles. They are the oldest and least expensive automatic detectors available. They also have the lowest rate of false alarms. However, they are fairly slow in detecting a fire in its initial stage and are better suited for small, confined spaces where high heat is expected.

Heat detectors can be either spot detectors, which are concentrated at a particular location, or continuous-line detectors, which are used mostly for cable trays and conveyors. There are three types of heat detector based on the way they operate: fixed temperature, rate compensation, and rate of rise.

Fixed-Temperature Type

As a spot detector, the fixed-temperature heat detector consists of two metals (each having a different coefficient of thermal expansion) that are bonded together. When heated, one metal will bend toward the one that expands at a slower rate, causing an electrical contact to close. This type of detector is very accurate and is set for various temperatures that can be expected to develop during a fire (e.g., 125, 140, 212, 275, and 350°F). It is also automatically self restoring, which means that after the operation is completed, the detector returns to its original shape or condition.

As a continuous-line detector, the fixed-temperature heat detector can use a pair of steel wires enclosed in a braided sheath to form a single cable (see Figure 4-1). The two concentric elements are separated by a heat-sensitive insulation. Under heat exposure, the insulation melts and the wires make contact. Since the portion affected must be replaced, this type is not self restoring. Another type of continuous-line, fixed-temperature heat detector uses two coaxial cables with temperature-sensitive semiconductor insulation between them. In case of high heat, the electrical resistance of the insulation decreases, and more current flows between the wires, caus-

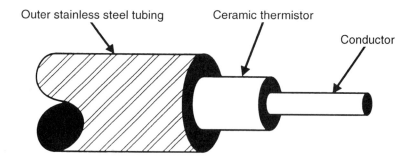

Figure 4-1 Continuous-Line, Fixed-Temperature Heat Detector

ing a contact to be initiated. This type of detection is self restoring because no insulation melting takes place during the process.

Rate-Compensation Type

The rate-compensation heat detector reacts to the temperature of the surroundings (see Figure 4-2). When the temperature reaches a predetermined level, regardless of the rate of temperature rise, an electrical contact is made. The difference between a rate-compensated detector and one with a fixed temperature is that the former eliminates the response at the peak temperature. The entire detector enclosure (rate compensation) must reach the critical (previously set) temperature and only then does it make contact, sounding an alarm or activating a fire-suppression system.

Rate-of-Rise Type

The rate-of-rise heat detector is effective when a rapid rise in temperature is expected due to a fire caused by a specific type of combustible (see Figure 4-3). This detector sounds an alarm and/or starts a suppression system when the temperature rise is faster than 15 to 25°F/minute. It will compensate for small fluctuations.

There are other types of heat detector available besides the ones mentioned here. One type of detector that is popular is a combination rate-of-rise/fixed-temperature model. The fire-protection technician should make an educated analysis when selecting a detector

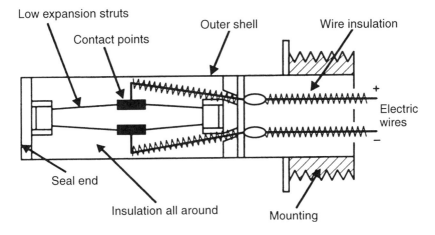

Figure 4-2 Rate-Compensation Heat Detector

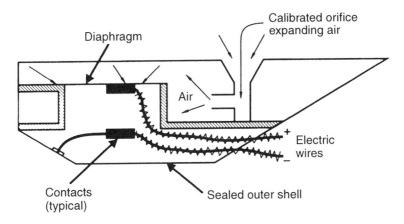

Figure 4-3 Rate-of-Rise Heat Detector

to do a job. The analysis should be based on all the detailed records, manufacturers' literature, and technical information available. Previous operational results should also be considered.

Smoke Detectors

Smoke detectors can be of either the ionization type or the photoelectric type. The photoelectric type is further divided into light-obscuration and light-scattering types.

Ionization Type

The ionization type is very common and uses a small quantity of low-grade radioactive material (e.g., americiu) to ionize the air within the detector and make it electrically conductive. If smoke enters the detector, the smoke particles attach themselves to the ions, and ion mobility is decreased. An alarm then sounds.

Photoelectric Type

In the photoelectric, light-obscuration type, the detector consists of a two-piece metal tube with a light source at one end and a receiving photo cell at the other (see Figure 4-4). Between the light source and the receiver is a light beam. The rising smoke from a fire obstructs the light normally traveling toward the receiving cell, which then causes the detector to sound an alarm. Special light filters prevent other light sources within the area from influencing the cell. This type is effective and has certain special applications due to the length of the light beam, which is operationally useful for a distance up to 300 linear ft.

The photoelectric, light-scattering type is similar to the light-obscuration type just described, except that the light and cell are located within the detector body and light beams do not normally fall on the receiving cell (see Figure 4-5). The light beam is scattered, so when the smoke rises, the light beam is redirected toward the receiving cell, which then makes a contact.

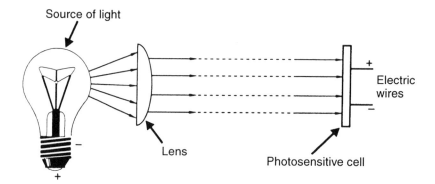

Figure 4-4 Photoelectric, Light-Obscuration Smoke Detector

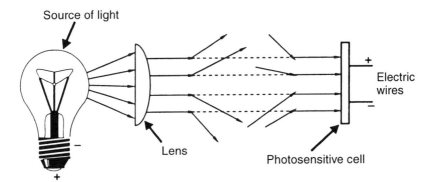

Figure 4-5. Photoelectric, Light-Scattering Smoke Detector

These are just two of the photoelectric types in which the receiving cell is independent of the emitting cell. There are other types with both cells in one enclosure.

Flame Detectors

Flame detectors respond to radiant energy and respond very quickly to a fire. They are often used in areas where there is a potential for an explosion.

Light is visible to the human eye when its wavelength is between 4000 and 7700 angstrom (Å).[1] When the wavelength is smaller than 4000 Å, it is ultraviolet light. When the wavelength is greater than 7700 Å, the designation is infrared light. Both types of light (ultraviolet and infrared) are invisible to the human eye. The ultraviolet light generated by the sun might produce false alarms, so detectors have been developed to reject sunlight and other unwanted radiation (e.g., from welding).

Lenses must be kept clean and free of dust or mist in order to be responsive and sensitive. One way to keep them clean is to provide an air shield. Compressed air is either blown over the lens, or a mechanism similar to windshield wipers on a car wipes the lens occasionally.

Infrared detectors operate best when they are separated from the flame by height and distance. They work well in large open areas where there is an accumulation of flammable liquids (e.g., aircraft hangars).

The sensing element is either a silicon solar cell or a sulfide cell made of lead or cadmium. A built-in time delay allows the detector to discern a flicker from a continuous infrared light emanating from a fire.

Choosing a Detector Device

Detector operational characteristics and physical location influence the selection of the detector type and its placement. Here are a few helpful guidelines to follow when selecting a detector:

- Combustion products—Certain detectors are sensitive to specific combustibles (e.g., flammable liquids or cellulose base materials) and no other products. The detector may only react if the smoke emanating from a material falls within certain parameters. For example, ionization type detectors may not detect large smoke particles, because they lack high mobility.
- Fire development—Fire development (speed) differs from oil fires to electrical fires to other kinds of fires. Some detectors will not detect all types of fire development.
- Ventilation—If a large ventilation air rate is normally needed for the area, then the combustion products may be drawn out of the area before they reach the detectors. This might be the case if the detector is mounted on the ceiling. The type of detector selected should be installed close to the area protected or close to the air exhaust from the room. The area surrounding the air supply might actually be kept free of smoke.
- Room congestion—Certain detectors have to "see" the fire. A maze of pipes, ducts, vessels, etc., may obstruct the hazard area.
- Room geometry—A very high room renders heat, photoelectric, and ionization detectors ineffective. The best choices for such an application are infrared or ultraviolet type detectors.
- Operational activities—Check whether the operational activities in the area may produce signals that would invol-

untarily trigger detector operation. For example, ionization detectors do not distinguish combustion products from a fire or from a diesel generator in operation. In a diesel generator room, heat detectors are recommended.

- Cost—If a large number of detectors are to be installed, the difference in equipment cost plus installation cost could become significant. Selecting the right detector is not an easy task. Experience gained with practice coupled with help from detector manufacturers and consultation with the authority having jurisdiction assist in finding the correct solution.

In addition to responding quickly to a fire, a good detection system should not sound any false alarms or initiate fire-suppression systems. Table 4-1 and Figure 4-6 give a summary of the different detector applications.

Detector Location and Spacing

Location and spacing of detectors must be consistent with the environment in which they operate and the qualifications for which they were tested. For example:

- Keep heat detectors away from normal heat sources such as space heaters. For spot heat detectors, it is best to install them on the ceiling or side wall (not closer than 4 in. from either). When the ceiling either does not have a smooth surface or has a height of more than 16 ft, the spacing is based on specific NFPA recommendations as well as the requirements of the authority having jurisdiction.
- Install smoke detectors closer to the return air register. They should not be installed close to the air supply into the area.
- Install flame detectors where they can "see" the fire.

For more information on fire-detection devices, see NFPA Standard no. 72.

Table 4-1 Detector Applications Summary

Type	Where to Use	Action	Recommended Use	Cost
HEAT DETECTORS				
Fixed temperature	Large open areas, to protect heat-generating equipment	Responds when a predetermined temperature is reached	Use limited to indoor applications; low false-alarm rate; a reliable device.	Low
Rate of rise	Large open areas	The rate-of-rise response to a specific temperature rise per minute	Should be used indoors; low false-alarm rate.	Low
Rate compensated	Large open areas, to protect heat-generating equipment	The detector and its enclosure has to reach a critical temperature. It compensates to spikes.	Should be used indoors; low false-alarm rate.	Low
SMOKE DETECTORS				
Photoelectric	Projected beam type used in open areas, high rack storage, computer rooms, and aircraft hangars	Smoldering fires	Must be used indoors.	Moderate
Ionization	Offices, computer rooms, combustible materials	Fast-flaming fires	Should be indoors	Moderate
FLAME DETECTORS				
Infrared	Hazardous work; explosive and rocket propellant manufacturing, aircraft hangars	Rapid response to infrared radiation generated by fire.	Indoor use; may be affected by heat	High
Ultraviolet	Hazardous work; explosive and rocket propellant manufacturing, aircraft hangars	Rapid response in milliseconds to ultraviolet radiation generated by fire.	May be used indoors or out; lenses need cleaning	High

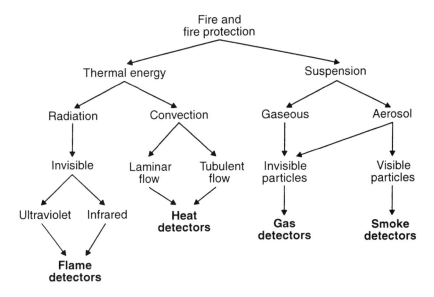

Figure 4-6 Detector Applications

Alarms

Like fire detectors, alarms do not fight fires directly. However, by alerting building occupants of a fire situation, alarms can save lives and/or property.

A fire-detection system is normally connected to an alarm system. NFPA Standard no. 13 requires the installation of local water-flow alarms in areas that have more than 20 sprinkler heads. This type of alarm provides a warning sound, which alerts personnel that water is flowing from one or more sprinkler heads. The alarm may be initiated by an alarm check valve installed in the system's riser. This check valve may be connected to an electric switch or a mechanical device, which activates a gong or bell.

Alarm systems are not detailed in this book, because specialized technicians in the electric/electronic field are responsible for the design and installation of such systems. However, alarm systems are always installed in cooperation with the fire-protection engineer or technician who establishes the criteria.

Notes

[1] An angstrom measures light wavelength and is equal to 1.0^{-10} meters.

Chapter 5:
Fire Suppression

In spite of fire-prevention methods, controls, and alarms, fires occur and endanger lives and property. For this reason, fire-suppression systems are necessary. These systems are comprised of various agents and methods, but whenever a fire starts, firefighters must be called.

The general strategy when fighting a fire is to locate it, surround it, confine it, and extinguish it. However, when firefighters arrive at the scene of a fire, their first concern is the safety of occupants who could be trapped.

When firefighters attack a fire in a conventionally low-height building, one of their first actions is to punch a hole in the building's roof so that heat and gases may escape. If confined, heat gases could hamper the firefighting capability and escalate the fire development.

Extinguishing Agents

Fire suppression involves an extinguishing agent and a means, system, or procedure to apply the extinguishing agent at the fire location. Extinguishing agents in an automatic suppression system should be selected based on several factors, including building construction material and building content. Other items that influence fire-suppression selection include:

- Type of combustible materials known or assumed to be involved in a fire in the protected area.
- Configuration of the area.
- Extinguisher expectations and performance.

- How the extinguisher affects one of the three elements involved in the fire triangle (see Figure 5-1).

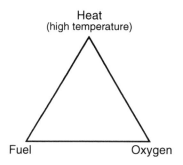

Figure 5-1 Fire Triangle

Extinguishing agents are selected based on suitability, cost, and cleanup required after a fire is out.

Table 5-1 shows the classification of combustible materials that may be involved in a fire and the type of suppression agent and application system recommended.

Table 5-1 Classification of Combustible Materials

Class	Combustible Materials	Suppression Systems and Agents
A	Wood, paper, cloth, rubber, insulated cables	Water (H_2O) sprinkler, spray; CO_2; foam designated as "Type A"
B	Flammable liquid, oil, grease	Foam, water (H_2O) spray; CO_2; dry chemicals
C	Electrical equipment	CO_2; Inergen (clean agent); H_2O
D	Combustible metals	Special agents

Note: It should be noted that in Table 5-1 insulated cables are considered Class A combustibles.

As stated previously, one of the goals of a fire-suppression system is to affect one of the three elements involved in a fire (oxygen, fuel, or heat). When fighting a fire that is either exposed to the atmosphere or involves an oxidizing agent, the goal is to lower the oxygen concentration below the minimum level (at or below 15% for general materials; 8% or lower for a smoldering, deep-seated fire in a cable tray) so combustion is not supported. One way to prevent the contact between the fire and the oxygen contained in the atmosphere is to apply a layer of inert gas over the fire surface in an enclosed space. If an area is unoccupied and can be leakproofed, "inerting" the respective room atmosphere is another possibility.

The temperature element of a fire may be controlled by cooling the combustion zone. The temperature should be lowered below the ignition temperature of the fuel vapors.

The most efficient cooling agent utilized in fire suppression is water, which is an extremely efficient heat absorber. Water is also inexpensive and available in most buildings through an existing network of pipes. Water is not dangerous or noxious to humans, and it can be cleaned up readily.

Regarding eliminating the fuel: Fires involving flammable gases are normally extinguished by cutting off the fuel supply at the source (such as closing a valve, which may be activated by a fusible link).

Water

Fixed water systems include hydrants on streets, hose stations or standpipe stations in buildings, and sprinklers in buildings. All of these systems require a reliable source of water supply and a connecting network of distribution pipes. The supply of water may come from the city water line or natural bodies of water such as rivers, lakes, or wells (fresh water only). *Note*: In areas with freezing temperatures, man-made reservoirs must be protected and checked daily.

A water source must be reliable. It must be available during droughts or freezing temperatures and able to supply the anticipated amount required as determined by the engineering calculations or available standards.

When the water-supply source cannot provide enough water flow, storage tanks may be installed to furnish the balance required during firefighting operations. NFPA no. 22 gives the standard installation and maintenance details for water tanks in private fire-protection systems.

The amount of water stored for fire-protection purposes varies with the type of hazard. Calculations take into consideration the standard amount of water stored as well as the flow required and the expected length of the suppression operation. These calculations determine a base storage requirement.

From the reservoir, water may be supplied to the extinguishing system by gravity (if the required head or pressure available is adequate) or with the help of pumps. The gravity system may be employed when the water source is located at an elevation high enough to provide the required working pressure at the sprinkler or hose station in the most remote location. When this pressure is not available, pumps are installed to deliver the flow capacity and pressure required for system operation.

If the supply system delivers a pressure that is lower than that required, booster pumps are installed. This type of pump boosts the pressure higher for proper system operation.

There are other fire-suppression systems available besides plain water. These include carbon dioxide (CO_2), halon or clean agents like Inergen, and dry chemical powders, which can smother a fire. Some chemicals may also be mixed with water for fire suppression. These systems are detailed in later chapters.

Chapter 6:
Fire Pumps

In a pressurized water-distribution system for fire protection, the first piece of equipment is the pump. The pump supplies and distributes water (through a network of pipes in the case of fire protection) from the source (reservoir or city water pipe) to the place of application. For the purpose of this book, a "pump" is defined as a mechanism that is used to push a liquid with a specific force to overcome system friction loss and any existing difference in elevation. The pump produces this force with the help of a motor or driver and consumes energy in this process.

Fire pumps are part of NFPA history. They were mentioned in the first standard issued in 1896, and in 1899 an NFPA committee was organized to study fire pumps. In the beginning, fire pumps were started manually. Modern fire pumps start automatically and are usually stopped manually.

It is imperative that pump selection be made carefully, so that it will work properly when operational. In addition to being sized properly, the pump must also be installed correctly and maintained regularly.

All fire pumps must be listed with UL (Underwriters' Laboratory).[1] The standard pump used for fire protection is a horizontal or vertical, centrifugal, single-stage or multistage pump. However, the pump can also be of the end-suction, in-line, horizontal split-case, or vertical-turbine design. Pump capacities range from 25 to 5000 gpm and pressures range from 40 to more than 500 pounds per square inch (psi). Electric motors (which do not have to be UL listed) and diesel drives (which must be UL listed) may occasionally exceed 500 hp. A special feature of a fire pump is the fact that it must deliver 150% of the rated capacity at no less than 65% of the rated head (pressure). In other words, a 1000 gpm pump rated at 100 psi must be capable of delivering 1500 gpm at a minimum of 65 psi. Another special feature is that the shut-off pressure of a fire pump (i.e., at zero capacity) must not exceed 140% of the pressure at the rated capacity. All fire pumps must be used with positive suction pressure, and they cannot be used for suction-lift applications. If a suction lift is required, a vertical-turbine pump must be used.

The capacity (Q) of a pump is the rate of fluid flow delivered, which is generally expressed in gallons per minute (gpm). The head or pressure furnished is the energy per unit weight of a liquid. The total head developed by a pump is the discharge head minus the suction (inlet) head:

$$H = h_d - h_s$$

where H = Total head, ft
h_d = Discharge head, ft
h_s = Suction inlet head, ft

Pump Components

The pump housing is referred to as the "casing." The casing encloses the impeller and collects the liquid being pumped. The liquid enters at the center or eye of the impeller. The impeller rotates, causing centrifugal force to push the liquid out (see Figure 6-1). The velocity is the greatest at the impeller's periphery, where the liquid is discharged through a spiral-shaped passage called the "volute." The shape is designed to provide an equal velocity of liquid at all circumference points.

The fire-pump assembly consists of a pump and a driver. Common drivers for fire pumps are electric motors, diesel engines, and steam turbines. The maximum speed of listed fire pumps is 3600 revolutions per minute (rpm).

There are pumps with double drivers (very rare) on the same shaft, electric at one end and diesel at the other. The most common driver is the electric-motor squirrel cage, induction type, three-phase, various voltage. Controllers are available for combined manual and automatic operation.

Figure 6-1 Impeller and Casing

Diesel drivers are not dependent on outside sources of power (electric or steam). A diesel driver is similar to a car engine, except that it is stationary and runs on diesel fuel oil (no. 2). A storage tank for no. 2 fuel oil should contain enough fuel for 8 hours of continuous pump operation and have a capacity of at least 1 gallon per horsepower plus a 10% safety factor. *Note*: 1 hp = 0.746 kW, or 3 kW = approximately 4 hp.

Diesel-engine controllers must have an alarm system to indicate:

- Low oil pressure (lubricating oil).
- Engineer jacket high coolant temperature.
- Failure to start automatically.
- Shutdown on overspeed.
- Battery failure.
- Battery charger failure.
- Engine running.
- Controller main switch turned from automatic to manual or off.

To ensure that the pump will start when required, it should have an optional timer, which will start the pump once a week and run it for a predetermined time (usually 15 minutes).

A few steps control a motor-driven, fire-pump operation:

- The pump should be primed.

- The circuit breaker shall be on.
- The pump shall start automatically in case of a drop in pressure in the system.
- The pump may be started manually (for test purposes).

Booster Pumps

As previously mentioned, when a fire-protection installation is supplied from a low-pressure water source, the system requires a booster pump. This type of pump raises the pressure in the water-supply line. For a relatively small installation, the pressure from the city water source is usually adequate.

The booster pump is selected based on the flow requirements and the pressure difference required. If, for example, the required operating pressure for a fire-protection system is 125 psi and the constant pressure available from the source (such as city water) is 50 psi, a booster pump is necessary. To calculate the booster-pump size required, find the difference between the required and available pressures, which in this case is 75 psi (125 psi – 50 psi). A safety factor of 10% should be added to the required pressure, so 125 psi + 12.5 psi (safety factor) – 50 psi = 87.5, or a 90 psi pump head selection.

Spare Pumps

In a large installation, spare pumps may be installed for emergency situations. The number of pumps to be installed depends on the situation. For example, if the total capacity required is 1500 gpm, two pumps could be installed, each with 1500 gpm at 100% capacity, with one pump being the spare. Alternatively, it would be possible to install three pumps, each at 50% capacity, or 750 gpm each. In this case, there is only 50% spare capacity, and for a limited fire, a 750 gpm pump starts instead of a 1500 gpm pump. The spare capacity is an added safety, which might be desired or requested by the authority having jurisdiction or the insurance underwriter. All pumps have the same design pressure.

Because there is no clear-cut solution to the question of spare pumps, every system must be analyzed independently. The final decision is usually made among the designer, owner, and authority having jurisdiction.

The fire-protection specialist should present the owner with available pump options, including pump type proposed, number of pumps, initial cost, maintenance requirements, and installation space required for each alternative. An educated decision can be made only after a detailed and specific analysis is performed.

Maintaining Pressure

In addition to a fire pump, a fire-protection installation includes a jockey pump or a hydropneumatic tank to maintain a constant, predetermined pressure in the sprinkler system and/or at the hose stations. A jockey pump may also compensate for minor leaks or a limited test of water discharge from the system.

Jockey Pump

The jockey pump is not a fire pump. It is a small pump with only 20 to 50 gpm capacity, but it has a discharge pressure (head) equal to the fire pump. It does not have the same special requirements as a fire pump.

Each fire-pump motor, jockey pump, or engine controller is equipped with a pressure switch. If the pressure in the system drops to a predetermined level, the jockey pump starts first. If the pressure in the system continues to drop because the flow cannot be satisfied, the fire pump starts.

The fire-pump system, when started by a pressure drop, should be arranged as follows:

1. The jockey-pump stop point should equal the pump work pressure plus the minimum static-supply pressure.
2. The jockey-pump start point should be at least 10 psi less than the jockey-pump stop point.
3. The fire-pump start point should be 5 psi less than the jockey-pump start point. Use 10 psi increments for each additional pump.
4. Where minimum run times are provided, the pump will continue to operate after attaining these pressures. The

final pressures should not exceed the pressure rating of
the system.

Example:
Pump: 1000 gpm, 100 psi pump with work pressure of 115 psi.
Suction supply: 50 psi from city—minimum static. 60 psi from
city—maximum static.

        Jockey-pump stop = 115 + 50 = 165 psi
        Jockey-pump start = 165 − 10 = 155 psi
        Fire-pump stop = 115 + 50 = 165 psi
        Fire-pump start = 155 − 5 = 150 psi
        Fire-pump maximum pressure = 115 + 60 = 175 psi

Hydropneumatic Tanks

Another way to maintain the water pressure in a sprinkler
system is to use a hydropneumatic tank. A hydropneumatic tank is
pressurized and consists of a small water-storage tank (100 to 200
gallons) with a cushion of compressed air in its upper portion (see
Figure 6-2).

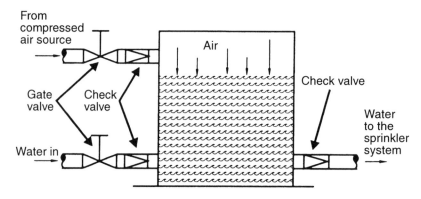

Figure 6-2 Hydropneumatic Tank

The volume of air and the tank pressure depend on whether the
hydropneumatic tank is located above or below the sprinkler heads.
If the tank is located above the sprinkler heads, the minimum pressure can be calculated as follows:

$$P = \frac{30}{A} - 15$$

where P = Air pressure, psi
         A = Volume of air in the tank (usually 33%, 50%, or 60%)

For example, if A = 0.33 (33%), the result is as follows:

$$P = \frac{30}{0.33} - 15 = 76 \text{ psi}$$

If the tank is located below the sprinkler heads, the minimum pressure can be calculated as follows:

$$P = \frac{30}{A} - 15 + \frac{0.434 + H}{A}$$

where P = Air pressure, psi
         A = Volume of air in the tank, %
         H = Height of the highest sprinkler head above the tank bottom, ft

The actual tank operating pressure is a function of the system pressure required. To determine the pressure in the tank when the system pressure is known, use the following calculation:

$$P_i = \frac{P_f + 15}{A} - 15$$

where P_i = Tank pressure, psi
         P_f = System pressure obtained from hydraulic calculations, psi

For example, if P_f = 75 psi and A = 0.5 (50%), the result is as follows:

$$P_i = \frac{75 + 15}{0.5} - 15 = 165 \text{ psi}$$

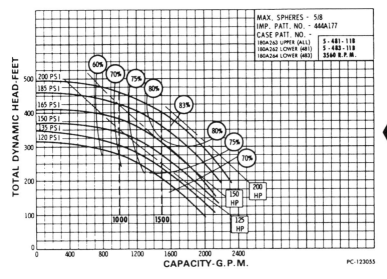

Figure 6-3 Pump Curve, 1000 gpm, 120 to 200 psi
(Courtesy, GS Aurora Pump)

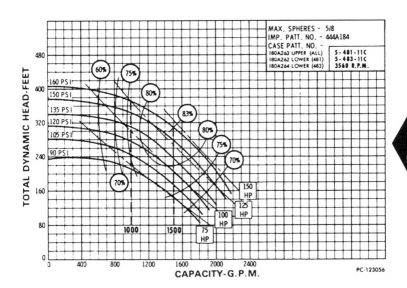

Figure 6-4 Pump Curve, 1000 gpm, 90 to 160 psi
(Courtesy, GS Aurora Pump)

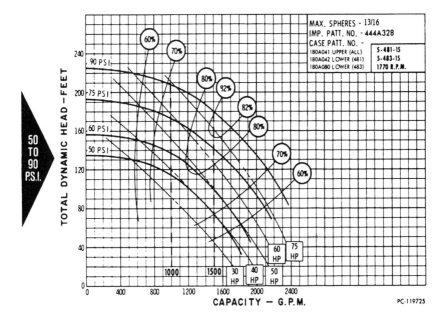

Figure 6-5 Pump Curve, 1000 gpm, 50 to 90 psi
(Courtesy, GS Aurora Pump)

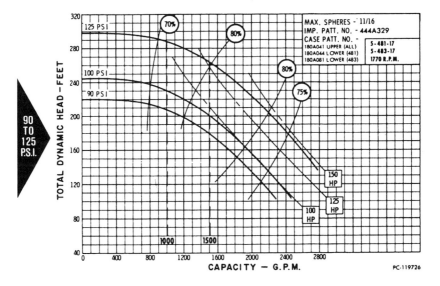

Figure 6-6 Pump Curve, 1000 gpm, 90 to 125 psi
(Courtesy, GS Aurora Pump)

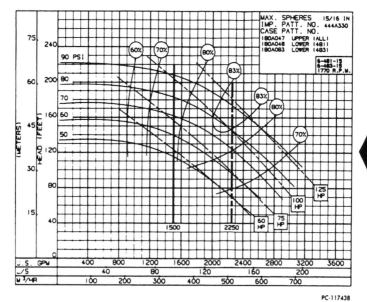

Figure 6-7 Pump Curve, 1500 gpm, 50 to 90 psi
(Courtesy, GS Aurora Pump)

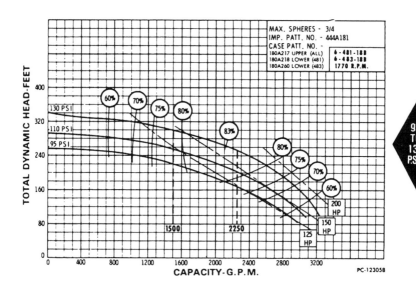

Figure 6-8 Pump Curve, 1500 gpm, 95 to 134 psi
(Courtesy, GS Aurora Pump)

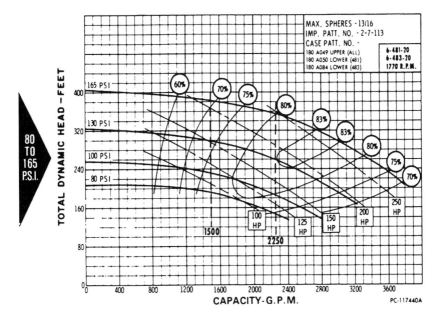

Figure 6-9 Pump Curve, 1500 gpm, 80 to 165 psi
(Courtesy, GS Aurora Pump)

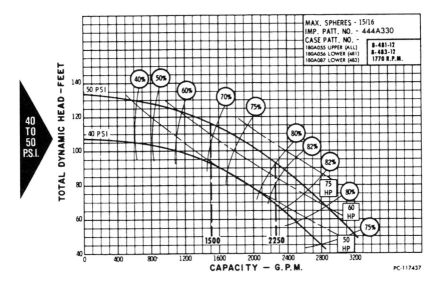

Figure 6-10 Pump Curve, 1500 gpm, 40 to 50 psi
(Courtesy, GS Aurora Pump)

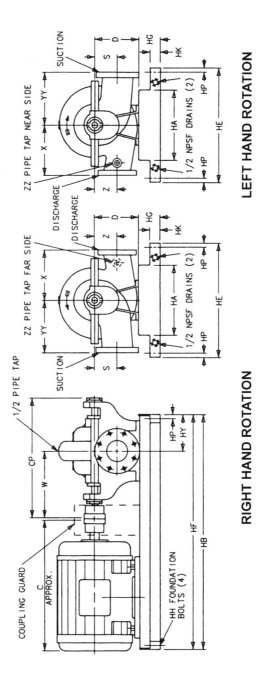

Figure 6-11 Pump Dimensions
(Courtesy, GS Aurora Pump)

A hydraulic calculation for a sprinkler system determines the amount of water and the head or pressure the pump must deliver and maintain for proper sprinkler-system operation. The pump selection is made based on flow and pressure.

Pump Curves

Figures 6-3 through 6-10 illustrate eight pump curves, four for 1000-gpm rated capacity pumps and four for 1500-gpm rated capacity pumps. Figure 6-11 illustrates the dimensions of a pump.

As mentioned previously, a fire pump must deliver 150% of the rated capacity at no less than 65% of the rated head (pressure). The pump curves indicate these conditions. For example, in Figure 6-4, when delivering 1500 gpm, following the 150 psi curve will work against a pressure of 280 ft of water, which represents 82%. This pump performs better than the code, which requires 65%.

Each pump curve diagram also includes the following information:

- Pump flow delivery capacity in gpm (horizontal line).
- Pump head or pressure capability measured in ft of water and/or the corresponding pressure in psi (vertical line).
- Required horsepower for electric motor (straight lines slanted down to the right).
- Pump efficiency (oval curves bending upward).
- Impeller rpm (written on the side).
- Range of pressure (written on the side).

Pump selection should be made for maximum efficiency, as this will save power when the pumps are running. Before making a final decision, discuss potential pump selections with a manufacturer's representative. This can be very helpful in selecting the proper pump.

In an installation, the fire pump must not be used for any other purpose. The fire-pump room should be kept at an ambient temperature (between 40 and 120°F), and it should be located on the ground floor. The fire department must have the ability to reach it quickly in case of a fire.

For more information on fire pumps, see NFPA Standard no. 20, *Installation of Centrifugal Fire Pumps.*

Notes

[1]This means equipment or materials used for fire protection are included in a list published by an organization acceptable to the authority having jurisdiction and concerned with equipment evaluation. This organization performs periodic inspections of listed equipment or materials. The listing states either that the equipment or material meets appropriate standards or has been tested and found suitable for use in a fire-protection application.

Chapter 7: Water-Suppression Systems (Manually Operated)

When a private water source is used (as opposed to a city water supply), facilities normally use a yard distribution system that includes a fire-protection piping loop. (The pipe diameter is calculated based on the demand and is usually 10 to 12 in. for a large installation.) This loop feeds the fire-protection system inside and provides outside protection through fire hydrants. It is equipped with sectionalizing, post indicator valves. These valves are important because if a portion of the loop does not function properly the valves will isolate that portion, allowing the rest of the system to operate normally.

Yard hydrants are located above the loop and approximately 50 ft away from the building. They are spaced approximately 250 ft apart. When required, the hydrants must be equipped with all the necessary accessories required for their operation. For example, if there are large amounts of combustibles in the yard (piles of lumber, oil storage), hydrants must have fixed nozzles permanently connected and aimed at the hazard area. Table 7-1 lists hydrant classes, colors, and flow capabilities according to NFPA Standard no. 291, *Hydrants Testing and Marking*. Please note that local rules might differ from this general rule.

Table 7-1 Hydrant Classification

Class	Top of Hydrant Color	Flow Capability
A	Green	1000 gpm or more
B	Orange	500 to 1000 gpm
C	Red	Less than 500 gpm

Hydrants must be operable all the time; therefore, they must be observed regularly for vandalism and other damages. They must also be lubricated on a yearly basis.

Standpipes

Standpipes provide a means of manual water application to a fire within a building. They are always needed when automatic systems are not provided, but they are often used even when an automatic system is installed. Standpipes are connected to water-supply mains or to fire pumps, tanks, and other equipment necessary to provide an adequate supply of water.

The installation of standpipes and hose systems is covered in NFPA Standard no. 14. The NFPA definition for a "standpipe system" is "an arrangement of piping, valves, hose connections, and allied equipment installed in a building or structure with the hose connections located in such a manner that water can be discharged in streams or spray patterns through attached hose and nozzles, for the purpose of extinguishing a fire and so protecting a building or structure and its contents in addition to protecting the occupants."

Hose-station systems and standpipes are grouped into three classifications:

- Class I: 2½-in. hose connection to be used by fire department and those trained to handle heavy streams
- Class II: 1½-in. hose connection to be used by building occupants
- Class III: 2½-in. hose connection with an easily removable adapter to a 1½-in. connection, which is intended for fire department or building occupants. An alternative, which is preferred by the fire department, is two parallel outlets on the same supply pipe, one 2½-in. and one 1½-in. diameter.

Table 7-2 illustrates these classifications.

Table 7-3 illustrates some of the data outlined in NFPA Standard no. 13. These data include water demand, length of operation expected for the hose stations, and the specified hazard. Table 7-3 is applicable only to sprinkler systems to which hose outlets are

Table 7-2 Standpipe Classifications

Class Diam.	Hose (in.)	Handler Hose	One Hose (gpm)	Each Additional Hose (gpm)	Total Flow (gpm)	Maximum Operation (min)	Residual Pressure (psi)
I	2½	Fire dept. or trained personnel	500	250	2500	30	100
II	1½	Bldg. occup.	100	100	—	—	65
III	2½ and 1½	Both of the above	500	250	2500	30	100

connected. (Standpipe outlets are usually connected to the same riser[s] as the sprinkler system; however, the local authority having jurisdiction may require standpipes to be on a separate riser[s] and have a separate supply pipe[s].)

Table 7-3 Hose-Stream Demand

Hazard Classification	Inside Hose (gpm)	Combined Inside and Outside Hose (gpm)	Duration (min)
Light	0, 50, or 100	100	30
Ordinary	0, 50, or 100	250	60–90
Extra hazard	0, 50, or 100	500	90–120

A standpipe system can be wet or dry. The wet type is full of water and under pressure at all times. When the hose valve is opened in a wet system, water comes out through the hose and its nozzle. The dry system has an automatic closed valve located in the main riser, which is opened by a manually operated remote-control device located at the hose station.

Table 7-4 illustrates the pipe schedule that may be used to size a standpipe system. It is also possible to use hydraulic calculations to size a system.

The applicable code and the authority having jurisdiction must be consulted for exact supply-pipe sizes, depending on the building height and the system used. Connections can be either male or

Table 7-4 Nominal Pipe Sizes for Standpipes and Supply Piping

Total Accumulated Flow (gpm)	Nominal Pipe Size (in.)		
	Total distance of piping from furthest outlet		
	<50 ft	50–100 lt	>100 ft
100	2	2½	3
101–500	4	4	6
501–750	5	5	6
751–1250	6	6	6
1251 and over	8	8	8

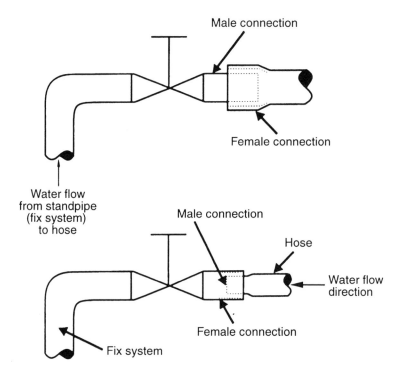

Figure 7-1 Female and Male Connections

female, depending on the direction of the water flow (see 7-1). However, connections must be determined in consu with the fire department.

Hoses

A "hose" is a rubber-lined, lightweight, linen-woven (or synthetic) jacket. Hoses may be stored on racks or reels with distinct markings that indicate their location.

While some local authorities may allow longer hoses, the maximum allowed length of a stored hose is usually 100 ft. This 100 ft of hose combined with a 30-ft water stream allows coverage of 130 ft from the hose station. For this reason, the distance between two hose stations (located in a straight line) is 150 ft for a building without sprinklers and 200 ft for a building with sprinklers.

The nozzle can deliver a solid stream of water or fog (with a special adapter). The fog-type nozzle is usually used when electric equipment and/ or combustible fluids are involved because, if directed on live wires, fog prevents the potential for operator electrocution. Fog also discourages the spread of fire. Some fire marshals do not allow the use of fog streams, so consult your local fire marshal first.

Chapter 8: Automatic Sprinkler Systems

Automatic sprinklers were developed to control, confine, and extinguish fires in order to prevent the loss of life and minimize the loss of property. However, the existence of a sprinkler system should not cause apathy among building owners and occupants. Flammable products, gases, liquids, or the accumulation of combustibles and other sensitive materials (e.g., explosives and rocket-propellant fuels) require strict supervision and continuous prevention and control.

A "fire-protection sprinkler system" is an integrated system of underground and overhead piping designed in accordance with fire-protection engineering standards. The design and installation of sprinkler systems is based on NFPA Standard no. 13, *Installation of Sprinkler Systems*. This standard was first published in 1896 and is the oldest NFPA standard. It was prepared in conjunction with:

- Fire-service personnel.
- Fire-insurance representatives.
- Testing laboratories for fire-protection items.
- Representatives from fire-protection equipment manufacturers.
- Contractors who installed such systems.

Due to continuous improvements made in automatic sprinkler systems, NFPA Standard no. 13 evolved into Standard no. 13A, *Inspection, Testing, and Maintenance of Sprinkler Systems*, in 1938. In 1960, NFPA's sprinkler committee redeveloped the sprinkler standard. As with any other code or standard, this standard gives only the minimum requirements in order to provide a reasonable degree of protection. Based on the owner's preference, additional protection may be installed for a higher degree of safety.

History of Sprinklers

The first sprinkler system in the United States was installed in 1852 and consisted of perforated pipe. The first automatic sprinkler was invented 12 years later. By 1895, sprinkler-system development was increasing, and the Boston area alone had nine different systems. Boston experienced the most significant growth in this area because of the number of hazardous textile mills in the vicinity.

Before 1950, sprinkler heads simultaneously discharged water upward and downward. The downward movement quenched the fire, while the upward movement kept the structure cool. These old-style heads were replaced by upright and pendent heads.

The current drive is to install automatic sprinkler systems even in residential buildings (NFPA Standards nos. 13D and 13R cover these applications.)

Selecting the Type of Sprinkler System

The factors to consider (or questions to be asked by the designer) in selecting the type of sprinkler system or the type of suppression system are as follows:

- Due to its content (combustible materials), is the area to be protected expected to develop a fast-growing fire?
- What is the principal goal of the fire-suppression system—occupants or content?
- Are there valuable items in the area protected that can be damaged by water?
- Is there a possibility of freezing?

Answering these basic questions will determine the type of suppression system to be designed and installed.

Sprinkler Definitions

There are various types of fixed-sprinkler system. Each system is clearly defined in NFPA Standard no. 13., and these definitions follow.

Wet-Pipe System—A system employing automatic sprinklers attached to a piping network containing water under pressure at all times and connected to a water supply so that water discharges immediately from sprinklers opened by a fire. Approximately 75% of the sprinkler systems in use are of the wet-pipe type. This type of sprinkler system is easy to maintain and is considered the most reliable. It is installed where there is no danger of freezing or special requirements.

Dry-Pipe Systems—A system employing automatic sprinklers attached to a piping system containing air or nitrogen under pressure, the release of which (as from a sprinkler opening) permits the water pressure to open a valve located in the riser known as a "dry-pipe valve." The water then flows into the piping system and out the opened sprinklers. A dry-pipe system starts somewhat more slowly than a wet one; however, the time between the sprinkler opening and the water flowing can be shortened by using quick-opening devices. This system is used where sprinklers are subject to freezing.

The dry-pipe system uses a general compressed-air system or a local air compressor. The air-supply line must have a restrictive orifice with a $\frac{1}{16}$-in. diameter. The sprinkler-head orifice must be larger than the supply-pipe opening or the air pressure will not drop and the dry valve will not open.

It should be emphasized that all components must be listed and approved.

Pre-Action System—A system employing automatic sprinklers that is attached to a piping system containing air that may or may not be under pressure, with a supplemental detection system installed in the same areas as the sprinklers. If the air is under pressure in the pipes, the pressure must be very low (just enough to help detect air leaks). Actuation of the detection system opens a valve, which permits water to flow into the sprinkler piping system and to be discharged from any sprinklers that may be open. This system is used where valuables are stored and accidental water discharge may cause damage.

Deluge System—A sprinkler system employing open heads attached to a piping system and connected to a water supply through a (deluge) valve, which is opened by the operation of a detection system installed in the same areas as the sprinklers. When this valve opens, water flows into the piping system and discharges from all heads attached thereto. This system is used in very high-hazard areas.

Combined Dry-Pipe and Pre-Action Sprinkler System— A system employing automatic sprinklers attached to a piping system containing air under pressure with a supplemental detection system installed in the same areas as the sprinklers. Operation of the detection system actuates tripping devices, which open dry-pipe valves simultaneously and without loss of air pressure in the system. Operation of the detection system also opens approved air-exhaust valves at the end of the feed main, which usually precedes the opening of sprinklers. The detection system also serves as an automatic fire-alarm system.

Antifreeze System—A wet-pipe system employing automatic sprinklers attached to a piping system that contains an antifreeze solution and is connected to a water supply. The antifreeze solution fills the pipes first, followed by water, which discharges immediately from sprinklers opened by heat from a fire. The antifreeze system is no different than a wet system except that the initial charge of water is mixed with antifreeze. The system may be installed in unheated areas as can a dry system. Additional devices may be required to prevent air-pocket formation. This system prevents the water from freezing in the pipes.

Sprinkler Operation

The sprinkler system inside a building is actually a network of pipes that are sized from either pipe schedules or hydraulic calculations. The system is installed overhead, and sprinkler heads are attached to the pipes in a systematic pattern. The valve controlling each system riser is located in the system riser or its supply piping.

Heat from a fire triggers the sprinkler system, causing one or more heads to open and discharge water only over the fire area (except in deluge systems with permanently open heads). Each sprinkler system includes a device for activating an alarm when water starts to flow.

Studies of more than 81,000 fires performed over a 44-year period indicate that sprinklers were effective in controlling 96.2% of the fires.[1] The automatic sprinkler system is a very reliable and efficient suppression system because of the following features:

- Immediate detection.
- The sounding of an alarm.
- Minimal response time.
- Continuous operation until the fire is completely extinguished.

Because sprinkler systems are so reliable, insurance companies reduce their rates considerably for buildings that are equipped with complete systems.

Care and Maintenance

Sprinkler heads shall never be stored where temperatures may exceed 100°F. Sprinkler heads shall never be painted, coated, or modified in any way after leaving the manufacturing premises. Care should be exercised to avoid the damage of sprinkler heads during handling

System Design

Fire-protection design documents consist of drawings and specifications. These documents must be prepared, approved, and kept readily available for further inspection and modifications if necessary. After installation, a fire-protection system must be inspected and tested.

When developing a sprinkler-system design, the code requires certain data to be listed on drawings. NFPA Standard no. 13 lists all the information required on the working design drawings, which includes the following:

- Name, location, address of property on which sprinklers will be installed.
- Owner and occupant.
- Point of compass (north direction).
- Type of construction.
- Distance from hydrant.
- Special hazard requirements, etc.

It is essential that sprinkler systems be designed to fit the particular hazard of a building or structure. NFPA Standards nos. 231 and 231C cover sprinkler systems for storage areas that require specific arrangements and specialized sprinklers.

Water Supply

An automatic sprinkler system should be connected to an automatic water-supply system, such as a municipal water main or an automatic fire pump. NFPA standards do not actually specify the type of automatic supply.

In addition to being reliable, the water supply must have the required pressure and capacity needed for the sprinkler system. The water quantity for the sprinklers is determined by adding the flow requirements for the number of sprinkler heads expected to operate plus 500 gpm for a hose station (unless otherwise directed by the authority having jurisdiction).

The number of sprinkler heads expected to operate in case of a fire depends on the following items:

- Occupancy.
- Combustibility of content.
- Height of stock pile.
- Combustibility of construction.
- Ceiling height.
- Horizontal and vertical cutoffs.
- Area shielded from proper water distribution.
- Type of sprinkler system used.

Strainers

Strainers are ordinarily required in sprinkler-system supply lines where the sprinkler head orifice is smaller than ⅜ in. They should have holes small enough to protect the smallest water passage in the nozzle or sprinkler head used. The use of galvanized piping downstream of the deluge valve is recommended in water-spray systems so spray nozzles will not become clogged by rust.

Piping

Figure 8-1 illustrates the different components in an indoor-sprinkler piping network. Each component is defined as follows:

System Riser—Above-ground supply pipe directly connected to the water supply.

Risers—Vertical pipes supplying the sprinkler system. All vertical pipes in a system are included in the definition of risers, with the exception of the system riser.

Feed Mains—Pipes supplying risers or cross mains.

Cross Mains—Pipes supplying the branch lines, either directly or through risers.

Branch Lines—Pipes in which the sprinklers are placed, either directly or through risers.

All valves and components used in a sprinkler system must be UL listed or approved.

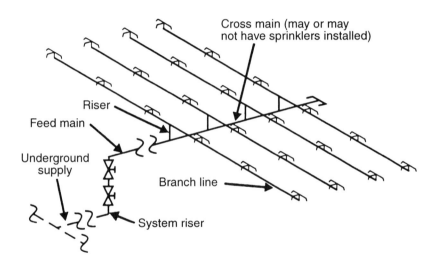

Figure 8-1 Indoor-Sprinkler Piping Network

Sprinkler systems for fire protection may also present public-safety risks. Preventing stale water from a fire-protection system from mixing with potable water is now a code requirement. Installing backflow preventers in fire-protection public supply or branches is a code requirement in most states and an essential component of fire-safety design and installation.

Pressure and Temperature

Sprinkler-system components are normally designed for a pressure of 175 psi, with a working pressure of 150 psi. Higher and lower design pressures may be used as required. If the pressure required in the system is higher than normal, then all system components must be rated for the higher pressure.

When the sprinkler system operating pressure is 150 psi or less, the test pressure must be 200 psi and the length of the test must be 2 hours. For any other operating pressure, the test must be the maximum operating pressure plus 50 psi. If the test takes place during the winter, air may be temporarily substituted for water.

Water temperature in a sprinkler system must be between 40 and 120°F. However, when water temperature exceeds 100°F, intermediate or higher-temperature sprinklers must be used.

Flushing

After installation, underground mains, lead-in connections, and risers must be flushed. This operation is very important, because factory-supplied pipes may contain dust, rust, etc., in addition to impurities collected during installation. If not eliminated, these foreign materials may block a sprinkler's orifice and render it inoperable. Table 8-1 shows the flushing rates prescribed by NFPA Standard no. 13.

Table 8-1 Flushing Rates

Pipe Size (in.)	Flow Rate (gpm)
4	400
6	750
8	1000
10	1500
12	2000

Source: NFPA Standard no. 13.

Area Limitation

The maximum floor area that may be protected by sprinklers supplied on each system riser on any one floor (as recommended by NFPA) is as follows:
- Light hazard: 52,000 ft^2 (4831 m^2)
- Ordinary hazard: 52,000 ft^2 (4831 m^2)
- High-piled storage: 40,000 ft^2 (3716 m^2)

- Extra Hazard:
  Pipe schedule—25,000 ft^2 (2323 m^2)
  Hydraulically calculated—40,000 ft^2 (3716 m^2)

System Drainage

All sprinkler systems must be installed so that the system may be drained if necessary. If repairs or alterations are required, a main drain valve will allow the system to be emptied. Wet-pipe systems may be installed level, while dry-pipe systems must be pitched for condensate drainage. The pitch is usually ½ in. per 10 ft for short branches and ¼ in. per 10 ft for mains. Mains must be pitched at least ½ in. per 10 ft in refrigerated areas.

Table 8-2 shows the recommended drain-pipe size as a function of the riser size. All valves and components should be UL listed or approved.

Table 8-2 Drain-Pipe Size

Pipe Size (in.)	Drain-Pipe Size (in.)
2 and smaller	¾ or larger
2½ to 3½	1¼ or larger
4 and larger	2

To determine the water-supply requirements for a pipe schedule, consult NFPA Standard no. 13, which gives flow rates and operational duration for light and ordinary hazards. Remember that the standard gives only minimum requirements. Better protection may be selected at an additional cost.

The use of pipe sizes based on a pipe schedule is somewhat restricted, so the designer must check applicable codes and standards. There is also a nomogram that indicates the water density in gpm vs square foot area that must be considered in the calculation of wet-type systems, depending upon the hazard type (see Chapter 10).

If the water used for domestic purposes is common with the one used for fire protection, a backflow preventer must be installed on the fire-protection line. Most state codes applicable to fire protection and/or plumbing make this installation mandatory in order

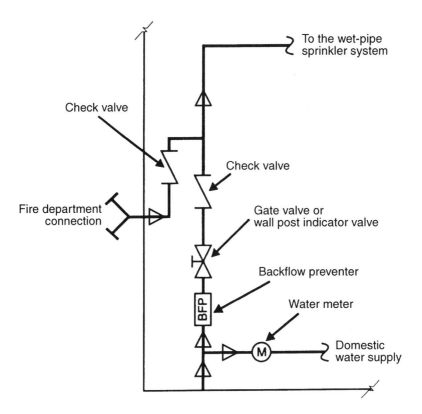

Figure 8-2 Backflow-Preventer Location

to protect the potable water from contamination. A schematic of the backflow-preventer location in the water supply is shown in Figure 8-2.

Sprinkler Components

There are many components in a sprinkler system, including:
- Single or multiple water supply.
- Piping underground and above ground connecting water supply to overhead sprinkler heads.[2]
- Fittings.
- Hangers and clamps.
- Associated hardware (control valves, check valves, alarm valves, dry-pipe valves, deluge valves, drain valve, and

pipe).
- Fire-department connections (Siamese connections).
- Alarms and annunciators.

While all these different components are vital to proper system operation, the sprinkler head is one of the most important components.

The automatic sprinkler head is a thermo-sensitive device that is automatically activated when an area reaches a predetermined temperature. Once this temperature is met, the sprinkler head releases a stream of water and distributes it in a specific pattern and quantity over a designated area. Water reaches the sprinklers through a network of overhead pipes, and the sprinklers are placed along the pipes at regular, geometric intervals.

Restraining Elements

Under normal conditions, water discharge from an automatic sprinkler head is restrained by a cap held tightly against the orifice. There are two types of restraining elements that are commonly used in sprinkler heads: fusible links and frangible bulbs.

In the fusible-link sprinkler head, a system of levers and links, which are anchored on the sprinkler frame, press on the cap to keep it firmly in place (see Figure 8-3). The system is constructed by fusing a metal alloy with a predetermined melting point. The metal is composed mainly of tin, lead, or cadmium (metals with low melting points). There are actually two different types of fusible link:
- *Solder-link type*—Constructed of a eutectic[3] alloy of tin, lead, cadmium, and bismuth. These metals have sharply defined melting points and, when alloyed in proper proportions, establish the operating temperature of the sprinkler.
- *Frangible-pellet type*—Has a pellet of either solder or another eutectic metal under compression, which melts at the design temperature and releases the cap.

The frangible-bulb restraining element is constructed of glass (see Figure 8-4). It is an enclosed bulb containing a colored liquid that does not completely fill the bulb. There is a small air bubble

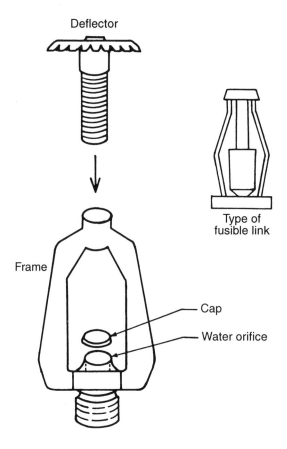

Deflector

Type of
fusible link

Frame

Cap

Water orifice

Figure 8-3 Fusible-Link Upright Sprinkler

entrapped in this colored liquid. When the temperature rises, the
liquid expands and the bubble is compressed and absorbed by the
liquid. As soon as the bubble disappears, the pressure in the bulb
rises rapidly and at a precise, preset temperature, the bulb shatters
and releases the cap. The exact operating temperature is regulated
by bubble size and the amount of liquid in the bulb. The higher the
operating temperature, the larger the bubble.

The recommended maximum room temperature is usually
closer to the operating temperature of the frangible-bulb type. This
is because in the fusible-link type, solder begins to lose its strength
below the actual melting point of the fusible link. If the duration of

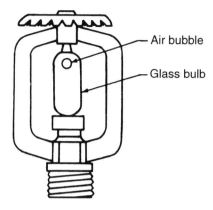

Air bubble

Glass bulb

Figure 8-4 Frangible-Bulb Upright Sprinkler

above-normal room temperature is excessive, premature sprinkler operation could occur.

Temperature Ratings

Sprinkler heads have various operating temperature ratings that are the result of standardized tests. The rating is stamped on the soldered link or restraining element. The frangible-bulb liquid color also indicates the sprinkler head temperature rating. Table 8-3 illustrates the temperature-rating color codes for fusible-link and frangible-bulb automatic sprinklers (with the exception of plated, flush, recessed, and concealed sprinkler heads) per NFPA Standard no. 13. The color is usually applied on the frame arms.

Table 8-3 Color Codes for Fusible-Link and Frangible-Bulb Sprinklers

Ceiling Temp. (°F)	Temp. Rating (°F)	Temp. Classification	Fusible-Link Color	Glass-Bulb Color
100	135 to 170	Ordinary	No color or black	Orange or red
150	175 to 225	Intermediate	White	Yellow or green
225	250 to 300	High	Blue	Blue
300	325 to 375	Extra high	Red	Purple
375	400 to 475	Extra high	Green	Black
475	500 to 575	Ultra high	Orange	Black

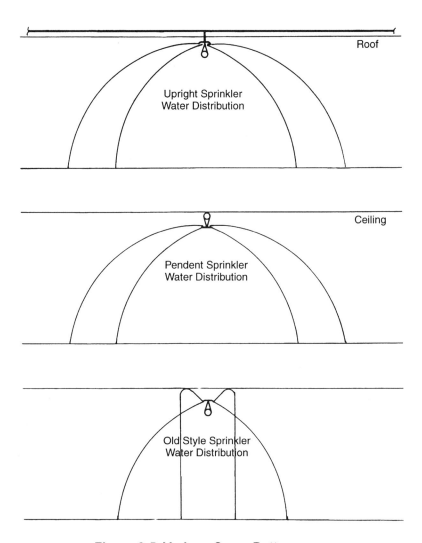

Figure 8-5 Various Spray Patterns

Deflectors

The deflector is attached to the sprinkler frame. When a water stream is directed against the deflector, it is converted into a spray of a certain shape designed to protect a defined area. The spray pattern depends on the deflector shape (see Figure 8-5). The pattern is roughly that of a half sphere filled with spray, in a relatively uniform distribution of water. For example, a spray may cover a circular area having a diameter of approximately 16 ft when the discharge rate is 15 gpm and the pressure is approximately 10 to 15 psi. In general, the gpm discharge is about 1.5 times the pressure required at the head (e.g., 15 psi and 22 gpm). *Note:* Do not use this for actual calculations.

The water discharge rate from a sprinkler head follows hydraulic laws and depends on the orifice size and water pressure. The standard sprinkler head has a ½-in.diameter orifice. Other orifice sizes can be easily identified by a protruding extension above the deflector. The orifice may be of the ring-nozzle or tapered-nozzle type.

Sprinkler-Head Types

Standard sprinkler heads are made for installation in an upright or pendent position and must be installed in the position for which they were constructed. Architects sometimes require special sprinkler types to be used for certain applications. There are over 20 types of commercially available sprinkler, including the following:

- *Upright*—Normally installed above the supply pipe.
- *Pendent*—Installed below the pipe.
- *Sidewall (horizontal and vertical)*—Similar to standard sprinkler heads except for a special deflector, which allows the discharge of water toward one side only in a pattern resembling one-quarter of a sphere. The forward horizontal range of about 15 ft is greater than that of a standard sprinkler. For special applications, a sidewall vertical type is used.
- *Extended coverage*—Covers more than 225 ft^2 per head.

- *Open*
- *Corrosion resistant*—Wax or Teflon coated by the manufacturer to protect against corrosives. Mostly regular pendent or upright type heads used in areas where corrosive substances are present (e.g., chlorine storage rooms and salt-water reservoirs).
- *Nozzles.*
- *Dry pendent and dry upright*—When a limited enclosure is subject to freezing, it may be connected to a wet-pipe system through a special dry-pipe connector.
- *Fast or quick response (QR).*
- *Quick response, extended coverage (QREC).*
- *Quick response, early suppression (QRES).*
- *Early suppression, fast response (ESFR).*
- *Ornamental.*
- *Recessed*—Most of the body is mounted within a recessed housing, and its operation is similar to a standard pendent sprinkler.
- *Flush*—Allows the working parts of the sprinkler head to extend below the ceiling into the area in which it is installed without affecting the heat sensitivity or the pattern of water distribution.
- *Concealed*—Entire body, including operating mechanism, is above a cover plate, which drops when a fire occurs, exposing thermo-sensitive assembly. Deflector may be fixed or it may drop below the ceiling level when water flows.
- *Old style.*
- *Residential.*
- *On-off sprinkler heads.*

Figures 8-6 through 8-12 illustrate some of these different types of sprinkler.

Water-Deluge Spray System

A water-deluge spray system refers to specially designed nozzles (open head) that force water into a predetermined pattern,

Figure 8-6 Upright Sprinkler
(Courtesy of The Viking Corporation)

Figure 8-7 Pendent Sprinkler
(Courtesy of The Viking Corporation)

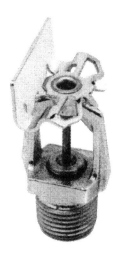

**Figure 8-8 Residential
Horizontal Sidewall Sprinkler**
(Courtesy of The Viking Corporation)

**Figure 8-9 Pendent Sprinkler
with Extra-Large Orifice**
(Courtesy of The Viking Corporation)

Figure 8-10 Upright with Large Drop Sprinkler
(Courtesy of The Viking Corporation)

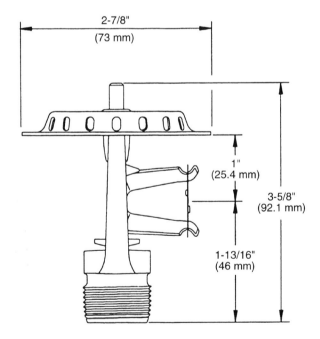

Figure 8-11 Sprinkler Dimensions

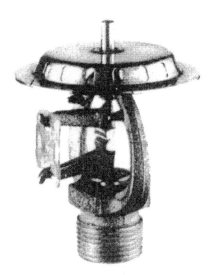

**Figure 8-12 Quick-Response, Specific-Application,
Upright Sprinkler**
(Courtesy of The Viking Corporation)

particle size, velocity, and/or density. Because a water-deluge system has all the nozzles initially open, there is heavy water consumption; therefore, each hazard should be protected by its own separate system (riser).

Piping and nozzle location, with respect to the surface or zone where the deluge system is applied, is influenced by physical arrangement. Other elements to be determined are the size of the nozzle orifice to be used (the flow required), the angle of the nozzle discharge sphere, and the required water pressure.

There are various shapes and sizes of nozzles, including high-velocity spray nozzles, which discharge in the form of a spray-filled cone, and low-velocity nozzles, which usually deliver a much finer spray in either a spray-filled sphere or cone shape. Another type of nozzle uses a slightly tilted deflector, and the angle of the spray discharge is governed by the design of the deflector.

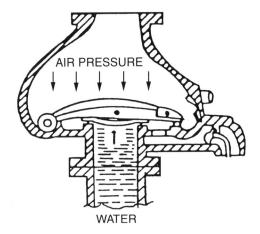

Figure 8-13 Deluge-Valve Schematic

Dry-Pipe or Deluge Valve

The dry-pipe valve or deluge valve is an important piece of equipment in dry-pipe and deluge systems. There are various types of valve, and while each dry-pipe valve type is of a slightly different design and construction, the operation is the same. This operation includes a clapper (check valve), which has special design features allowing air pressure on one side of the valve while the opposite side restrains higher water pressure (see Figure 8-13). When a fire occurs and sprinkler heads open, the air pressure drops within the valve, opening the clapper and filling the pipes. The valve trips an alarm when water flows through the alarm connections.

Another type of deluge valve involves the water-supply pressure exerting pressure on the clapper in the diaphragm chamber. When the activation devices operate, water from the diaphragm chamber is released faster than it can be replenished, destroying the pressure equilibrium and allowing water to flow into the piping system.

Deluge valves are usually equipped with various trim arrangements for manual and/or automatic activation by an electrical signal, which can be operated either pneumatically or hydraulically.

It is up to the design engineer and the owner to make the selection of the trim arrangements.

For areas where water damage and/or consumption is of great concern, manufacturers have built a recyclable type of deluge valve that permits automatic, remote on-off control. The valve opens when a fire occurs and automatically shuts the water off when the heat is reduced below the detector operating temperature. It has the capability of turning the water back on when the set temperature is exceeded again. All other valves must be closed manually.

Sprinkler Installation

It is critical to determine proper sprinkler location before installing any system. The following points must be taken into consideration before a system is installed:

- Maximum protection area per sprinkler head.
- Minimum interference to the discharge patterns by structural elements, piping, ducts, or lighting fixtures.
- Correct location with respect to structural elements to obtain suitable sensitivity to potential fires.

In general, the maximum distance between sprinklers on branches for light and ordinary hazard occupancies is 15 ft. The protected area coverage per sprinkler head, as required by NFPA Standard no. 13, is as follows:

- *Light-hazard occupancy*: 168 to 225 ft^2 depending on construction type.
- *Ordinary-hazard occupancy*: 100 to 130 ft^2.
- *Extra-hazard occupancy*: 90 to 100 ft^2.

All codes require sprinkler systems to have devices that will sound an alarm when water flows through the risers or main supply due to a fire, accidental rupture of piping, or head(s) opening. These devices also monitor all valves to ensure that they are in the correct operating position. This can be achieved by a remote signaling to a control station, sounding an alarm, or locking the valve in an open position. In other words, the devices supervise the system and sound an alarm when any tampering or undesired/unnecessary operation is detected.

Only UL listed materials and equipment may be used in sprinkler installations. In addition, sprinkler heads must be installed in accordance with their listing, and sprinklers must not be altered (painted or any coat of protective material applied in the field or at the job site). It is very easy to determine whether a sprinkler head is painted by the manufacturer or in the field; if painted by the manufacturer, the operative parts are left unpainted.

When sprinkler heads must be replaced, the same type must be used. This means the same orifice type and temperature rating, unless there are new conditions, such as a change of occupancy or structural modifications (e.g., added or canceled ceiling).

Alarms

Three basic types of alarm can be part of a sprinkler system:

1. *Vane-type water flow*—Comes equipped with a small paddle that is inserted directly into the riser pipe. The paddle responds to water flow as low as 10 gpm, which then triggers an alarm. This type may be equipped with a delayed system (adjustable from 0 to 120 seconds) to prevent false alarms caused by normal water-pressure fluctuations.

2. *Mechanical water-flow alarm (water motor gong)*—Involves a check valve that lifts from its seat when water flows. The check valve may vary as follows: (a) Differential type has a seat ring with a concentric groove connected by a pipe to the alarm device. When the clapper of the alarm valve rises to allow water to flow to the sprinklers, water enters the groove and flows to the alarm-giving device. (b) Another type has an extension arm connected to a small auxiliary pilot valve, which, in turn, is connected to the alarm system.

3. *Pressure-activated alarm switch*—Used in conjunction with dry-pipe valves, alarm check valves, and other types of water-control valve. It has contact elements arranged to open or close an electric circuit when subjected to increased or reduced pressure. In most cases, the motion to

activate a switch is given from a diaphragm exposed to pressure on one side and supported by an adjustable spring on the other side.

The alarm for a dry-pipe sprinkler system is arranged with a connection from the intermediate chamber of a dry-pipe valve to a pressure-operated alarm device. When the dry-pipe valve trips, the intermediate chamber, which normally contains air at atmospheric pressure, fills with water at the supply pressure, which operates the alarm devices. Sometimes both an outdoor water motor gong and a pressure-operated electric switch are provided. The alarm devices for the deluge and pre-action systems are of the same type as those used for the dry-pipe system.

Codes require water-supply control valves to indicate conditions that could prevent the unwanted or unnecessary operation of the sprinkler system. This can be achieved by using electric switches, also called "temper switches," which can be selected for open or closed contact. The signal that indicates valve operation is given when the valve wheel is given two turns from the wide open position. The restoration signal sounds when the valve is restored to its fully open position. This simply cancels the temper-switch alarm.

Notes

[1]In cases where the sprinklers were ineffective, studies show the reasons for failure include: improper water supply or system was not adequate, valve was in the wrong position (closed instead of open), and system was taken out of operation without temporary replacements.

[2]The life line of a sprinkler system is the distribution-piping network, which conveys the agent to the fire. It must be the correct size, well constructed, and well supported.

[3]An alloy having the lowest melting point possible, which means lower than each of the components.

Chapter 9: Hydraulic Principles and Properties of Water

"Hydraulics" is the study of fluid behavior when fluid is either resting or flowing. From a fire-protection point of view, the focus is on water flow in pipes.

Water normally flows in pipes by gravity from a higher to a lower elevation. If flow is in the opposite direction, water is always helped along by a pump (see Chapter 6), which consumes energy.

"Static head" is the amount of potential energy due to the elevation of water above a certain reference point. Static head is measured in feet (ft) of water and can be converted into pounds per square inch (psi). Another element used in calculations is the "velocity head," which is the pressure loss in a system due to the velocity of water flow. These different elements are used in the examples of hydraulic calculations later in this chapter.

Besides the pipes, a water-distribution system includes fittings and valves. Fittings include elbows (standard 30, 45, or 90°, standard or long sweep), tees, crosses, sudden enlargements, or contractions, etc. Valves include control valves (globe, gate, angle, butterfly) and check valves (swing, lift, spring-loaded, water-type, etc.). Water flowing through pipes, fittings, valves, and equipment experiences friction, or pressure loss. This happens due to the "rubbing" of water molecules against the walls as well as additional turbulence that occurs while water flows through fittings and valves.

Water used for firefighting does not have to be potable. It can be raw, untreated water that is strained to eliminate impurities. However, it should be fresh water, as sea (salt) water is very corrosive. Systems using sea water are normally dry and must be constructed of noncorrosive materials, such as stainless steel, which is very expensive.

Protection of Water-Source Quality

The water supply for a firefighting system for a building usually originates at the public (potable) water system.

It is possible that the fire-protection water network will be separate from the potable water. Upstream at the reservoir the supply is common. Such a system applies in cities and towns. In the country, water is supplied from wells and stored in elevated tanks.

The water in a wet-sprinkler/hose-stations system is stagnant except during the occurrence of a fire. For this reason, the fix fire-protection system installed in a building should be equipped with a device to prevent the stagnant (possibly contaminated) water from returning into the potable-water system. This device is a "backflow preventer" of the double-check valve type.

Potential contamination should be controlled at any cross connection between fire-protection water and potable water. The potable-water system must remain safe at all times. The double-check valve prevents over-pressure in the fire-protection system (from the fire pump) from pushing water back into a sometimes lower public-supply system.

The federal Safe Drinking Water Act, passed in 1974 and expanded in 1983, mandates such protection. There are also local codes and rules based on American Water Works Association (AWWA) Manual no. 14, *Recommended Practice for Backflow Prevention and Cross-Connection Control.*

Properties of Water

The natural chemical properties of water (H_2O) are as follows: solubility (it dissolves a variety of substances), hardness, specific electrical conductance, hydrogen-ion concentration (pH, which at the value of 7 shows that the water is neutral, neither alkaline nor acidic), dissolved carbon dioxide, and dissolved solids. Some of the physical properties of water include: density, viscosity, compressibility (water is noncompressible), its boiling point, and its freezing point. It is important to recognize these properties, because in one way or another, they influence the flow of water in a system.

Density

By definition, "density" is the ratio of the mass (weight) of a substance to the volume it occupies. Density is measured in pounds per cubic foot, which can be written as "lb/ft³" or "lb/cu ft." In plumbing calculations, water density is usually considered to be 62.3 lb/ft³. This value represents the density of water at a temperature of 70°F (room temperature). Water density varies slightly with the temperature; the warmer the water, the less dense it becomes.

Viscosity

Viscosity concerns the friction of water molecules among themselves, as well as along the walls of the pipes and fittings. It is the physical property that directly influences the flow of water in pipes. The forces at work between the water molecules themselves are called "cohesion" and "adhesion." These forces can be measured in the laboratory.

Cold water is more cohesive than warm water, thus its viscosity is greater. This greater viscosity increases the friction of the flow of cold water through pipes. Warm water flows somewhat more easily through pipes because it is not as cohesive. However, the actual difference in viscosity between domestic cold and hot water is so small that it is considered insignificant for practical purposes and is negligible in calculations.

Viscosity is measured in centistokes or centipoise. At 60°F, water has a viscosity equal to 1.31 centistokes, which corresponds to a measurement of 0.00001216 ft²/sec.

Boiling Point

Water boils at 212°F (100°C) at sea level (atmospheric pressure). If the pressure varies, the boiling temperature point will also vary. The lower the pressure exerted upon the surface, the lower is the boiling point. For example, atmospheric pressure is lower on top of a mountain, so water boils at a lower temperature. The changes in the boiling point as a function of pressure are shown below:

Absolute Pressure (psi)	Water Boiling Point (°F)
1	101.8
6	170.1
14.7 (atmospheric)	212.0

Water Flow

The flow of water can be characterized as "laminar" or "turbulent." In laminar flow, streams of water molecules flow naturally parallel to each other up to a certain velocity. Above that velocity, the flow becomes turbulent. This characteristic was demonstrated by Osborne Reynolds, who developed a simple formula to determine the Reynolds number (R), which classifies the flow as laminar or turbulent. If R is less than 2000, the flow is laminar. The formula is as follows:

$$R = \frac{VD}{v}$$

where V = Water velocity in pipe, ft/sec
         D = Pipe diameter, ft
         v = Viscosity, ft²/sec

The velocity of water is greatest at the center of a pipe. More friction exists along the walls, where water molecules rub against pipe walls (see Figure 9-1). For the purposes of the problems given in this chapter, an average velocity represents 80% of the maximum velocity at the center of the pipe.

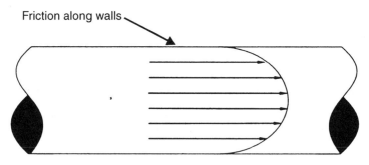

Friction along walls

Figure 9-1 Flow of Water in Pipes

Water velocity in pipes used for fire-protection applications normally ranges from 8 to 22 feet per second (fps). Conservative estimates in pipe size calculations recommend a maximum limit of 16 to 18 fps.

The basic formula for fluid flow is:

$$Q = AV$$

where   Q = Flow, ft^3/sec
A = Cross section of the pipe, ft^2
V = Water velocity in pipe, ft/sec

Another item often used in plumbing calculations is the "velocity head." This is defined as the decrease in head (or the loss of pressure), which corresponds to the velocity of flow. The formula for velocity head is:

$$hv = \frac{V^2}{2g}$$

where   hv = Velocity head, ft
V = Velocity in pipe, ft/sec
g = Acceleration of gravity, ft/sec^2

The formula to calculate the water velocity when velocity and pipe diameter are known is:

$$V = \frac{0.4085Q}{d^2}$$

where   V = Water velocity in pipe, ft/sec
Q = Flow, gpm
d = Pipe diameter, in.
0.4085 = Coefficient

Hydraulics

Hydraulics is part of a larger branch of physics called "fluid mechanics." Hydraulic principles are based on the chemical, physical, and mechanical properties of water. These properties were discussed earlier and include density, viscosity, type of flow (laminar or turbulent) as a function of velocity, water temperature, and pressure.

A piping system or network includes pipes, fittings, and valves. As discussed earlier, water flowing through these pipes, fittings, valves, and other equipment produces friction, or a loss of pressure. This pressure loss or resistance to flow occurs because the molecules of water "rub" against the walls of the pipes and fittings. The types of energy involved in the flow of water include kinetic and potential energy.

Calculations

Calculations are important to ensure that pipes are sized correctly. If the pipes are not sized correctly or the flow of water stops suddenly, the dynamic force of the flow may produce water hammer, shock, or noise. If the velocity is constantly too high when water is flowing, erosion may occur in the pipes.

When performing these calculations, consideration needs to be given to factors that affect the flow of water through the pipes. These factors include static head and friction or pressure losses that occur when water flows through pipe, pipe fittings, or equipment (e.g., water meters or heaters) (see Figure 9-2).

Measurements

The following is a short list of some units of measurement that are useful in solving the problems that follow:

$$\text{Acceleration of gravity} = 32.2 \text{ ft/sec}^2$$
$$1 \text{ ft}^3 \text{ H}_2\text{O} = 62.3 \text{ lb}$$
$$1 \text{ gal} = 0.1337 \text{ ft}^3$$
$$1 \text{ gal H}_2\text{O} = 8.33 \text{ lb (at } 70°\text{F)}$$
$$1 \text{ ft}^3 = 7.48 \text{ gal}$$

Therefore:

$$1 \text{ ft}^3 = (7.48)(8.33) = 62.3 \text{ lb}$$
$$1 \text{ atm} = 14.696 \approx 14.7 \text{ lb/in.}^2, \text{ or psi}$$
$$= 29.92 \text{ in. Hg (mercury)}$$
$$1 \text{ atm} = 33.96 \text{ ft of water}$$
$$1 \text{ ft of H}_2\text{O} \times 0.433 = 1 \text{ psi (see footnote #1)}$$

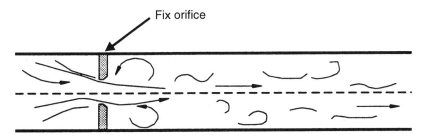

A. How an orifice influences the flow in pipe

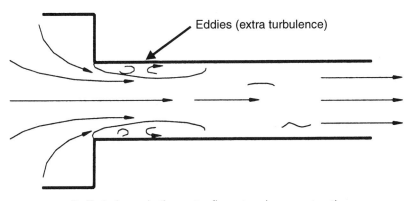

B. Turbulence in the water flow at a sharp contraction

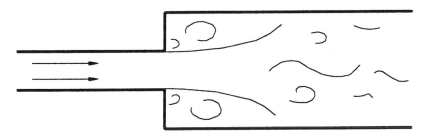

C. Tubulence in the water flow at a sudden expansion

Figure 9-2 Friction in Fittings

Examples

Example 9-1 What is the outlet pressure for water flow in the following straight pipe?

Pipe material: Schedule 40, standard-weight steel
Pipe diameter: 2 in.
Flow: 40 gpm
Length of pipe: 50 ft
Inlet water pressure 25 ft = 10.82 psi

Solution 9-1 Remember that when water flows along the pipe, friction, loss of head, or loss of pressure result. Also realize that while data is provided for calculations in this book, in an actual application, this data must either be calculated (e.g., required flow) or selected (e.g., piping material and limiting water velocity in pipes).

Based on the information that can be obtained from the tables in Appendix B, at a flow of 40 gpm, the velocity of water is 3.82 ft/sec and the head loss is 3.06 ft per 100 ft of pipe (Schedule 40 steel). The pipe length in this problem is 50 ft, so the head loss is as follows:

$$(50 \text{ ft}) \left(\frac{3.06 \text{ ft}}{100 \text{ ft}}\right) = 1.53 \text{ ft}$$

The pressure available at the pipe exit is now:

$$25 \text{ ft} - 1.53 \text{ ft} = 23.47 \text{ ft, or } 10.16 \text{ psi}$$

(See Figure 9-3.)

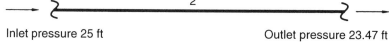

Inlet pressure 25 ft   Outlet pressure 23.47 ft

Figure 9-3   Straight Pipe

Example 9-2 Based on the data given in Example 9-1, what is the outlet pressure for water flow in pipe with these fittings: two 90°

standard elbows and a gate valve (see Figure 9-4)? (There is no difference in elevation.)

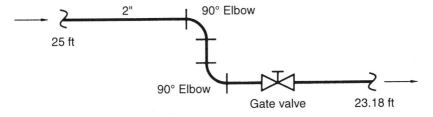

Figure 9-4   Pipe with Fittings

Solution 9-2   Remember, the total friction in the system consists of friction in the pipe plus the friction in fittings and valve.

The velocity head, which may be found in Appendix B, is calculated as follows:

$$hv = \frac{V^2}{2g} = \frac{3.82^2}{(2)(32.17)} = 0.227$$

The friction in fittings (hf) can be found in Appendix B and is based on the velocity head. The K value is a resistance coefficient, which helps in the calculation of the friction in fittings. It represents the coefficients for each, individual fitting, the sum of which equals total friction in fittings (e.g., K1 + K2 + ..., where K represents each fitting). For the problem at hand, the K value for a 2-in., 90°, standard elbow is 0.57. The K value for a gate valve is 0.15. Thus, the friction loss in fittings is as follows:

$$hf = [K1 \ (elbow) + K2 \ (elbow) + K3 \ (valve)] \left(\frac{V^2}{2g}\right)$$

$$= (0.57 + 0.57 + 0.15) \ (0.227) = 0.293 \ ft$$

We already calculated the friction in this length of pipe (see Example 9-1), so just add it to the friction in fittings and valve:

$$1.53 \ ft + 0.293 \ ft = 1.823 \ ft$$

The pressure available at the pipe exit is now:

$$25 \ ft - 1.823 \ ft = 23.177 \ ft, \ or \ 10.03 \ psi$$

In this example, the available pressure is lower because of the added friction in the fittings and valve.

Example 9-3 Based on the data given in Examples 9-1 and 9-2, what is the outlet pressure for water flow in pipe with a vertical portion (see Figure 9-5)?

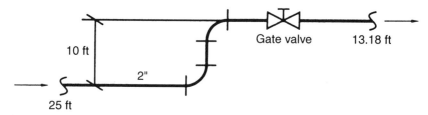

Figure 9-5 Pipe with Vertical Portion
(difference in elevation between the two pipe ends)

Solution 9-3 The calculations made in Example 9-2 remain unchanged since the data is the same. However, there is now some pressure loss due to a 10-ft vertical section.

The friction in pipe and fittings remains the same, at 1.823 ft. From the inlet pressure, in addition to fittings and valve, deduct the difference in elevation:

$$25 \text{ ft} - 1.823 \text{ ft} - 10 \text{ ft} = 13.177 \text{ ft, or } 5.70 \text{ psi}$$

The following examples are more detailed than those given previously. The calculations for these examples are similar to those for many hydraulic problems.

Example 9-4 Water must be transferred from a holding tank located at a low elevation to a reservoir located at a higher elevation (see Figure 9-6). Calculate the total head loss in the system in order to select the appropriate pump size to do the job. Use the following data for this calculation:

Pipe material: Schedule 40, standard-weight steel
Pipe diameter: 3 in.
Flow: 130 gpm

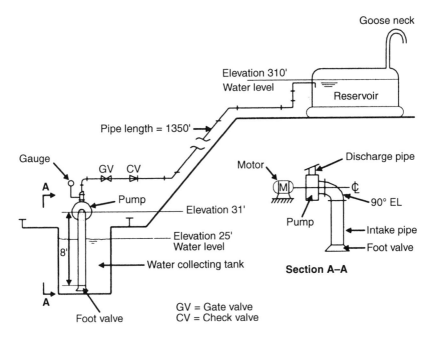

Figure 9-6 Transfer of Water from a Lower Elevation to a Higher Elevation

Water velocity: 5.64 ft/sec (see Table 9-1)
Friction: 3.9 ft per 100 ft of pipe (see Table 9-1)

Solution 9-4 The system shown in Figure 9-6 can be divided into two sections: the suction side of the pump and the discharge side of the pump. The head loss in the suction side of the pump includes: friction loss in the pipe, friction loss in fittings and valve, and static head (since the flow is against gravity, it is considered a loss). Static head must be added to the friction loss, since the pump must overcome both of these factors to push the column of water up through the pipes.

The friction loss in the pipe can be calculated with Darcy's formula, which is discussed later in this chapter. However, for purposes of this calculation, the corresponding values will be taken directly from Table 9-1.

Table 9-1 Friction of Water in 3-In. Pipe*

Asphalt-Dipped Cast Iron and New Steel Pipe (Based on Darcy's Formula)

Flow, U.S. gal per min	Asphalt-Dipped Cast Iron 3.0-in. inside dia.			Standard-Wt. Steel–Sch. 40 3.068-in. inside dia.			Extra-Strong Steel–Sch. 80 2.900-in. inside dia.			Schedule 160 Steel 2.624-in. inside dia.		
	Velocity, ft per sec	Velocity, head-ft	Head loss, ft per 100 ft	Velocity, ft per sec	Velocity, head-ft	Head loss, ft per 100 ft	Velocity, ft per sec	Velocity, head-ft	Head loss, ft per 100 ft	Velocity, ft per sec	Velocity, head-ft	Head loss, ft per 100 ft
10	0.454	0.000	0.042	0.434	0.003	0.038	0.49	0.00	0.050	0.593	0.005	0.080
15	0.681	0.010	0.088	0.651	0.007	0.077	0.73	0.01	0.101	0.89	0.012	0.164
20	0.908	0.010	0.149	0.868	0.012	0.129	0.97	0.02	0.169	1.19	0.022	0.275
25	1.13	0.02	0.225	1.09	0.018	0.192	1.21	0.02	0.253	1.48	0.034	0.411
30	1.36	0.03	0.316	1.3	0.026	0.267	1.45	0.03	0.351	1.78	0.049	0.572
35	1.59	0.04	0.421	1.52	0.036	0.353	1.70	0.04	0.464	2.08	0.067	0.757
40	1.82	0.05	0.541	1.74	0.047	0.449	1.94	0.06	0.592	2.37	0.087	0.933
45	2.04	0.06	0.676	1.95	0.059	0.557	2.18	0.07	0.734	2.67	0.111	1.16
50	2.27	0.08	0.825	2.17	0.073	0.676	2.43	0.09	0.86	2.97	0.137	1.41
55	2.50	0.10	0.990	2.39	0.089	0.776	2.67	0.11	1.03	3.26	0.165	1.69
60	2.72	0.12	1.17	2.6	0.105	0.912	2.91	0.130	1.21	3.56	0.197	1.99
65	2.95	0.14	1.36	2.82	0.124	1.06	3.16	0.15	1.4	3.86	0.231	2.31
70	3.18	0.16	1.57	3.04	0.143	1.22	3.40	0.18	1.61	4.15	0.268	2.65
75	3.40	0.18	1.79	3.25	0.165	1.38	3.64	0.21	1.83	4.45	0.307	3.02
80	3.63	0.21	2.03	3.47	0.187	1.56	3.88	0.23	2.07	4.75	0.35	3.41
85	3.86	0.23	2.28	3.69	0.211	1.75	4.12	0.26	2.31	5.04	0.395	3.83
90	4.08	0.26	2.55	3.91	0.237	1.95	4.37	0.29	2.58	5.34	0.443	4.27
95	4.31	0.29	2.83	4.12	0.264	2.16	4.61	0.33	2.86	5.63	0.493	4.73
100	4.54	0.32	3.12	4.34	0.293	2.37	4.85	0.36	3.15	5.93	0.546	5.21
110	4.99	0.39	3.75	4.77	0.354	2.84	5.33	0.44	3.77	6.53	0.661	6.25
120	5.45	0.46	4.45	5.21	0.421	3.35	5.81	0.52	4.45	7.12	0.787	7.38
130	5.90	0.54	5.19	5.64	0.495	3.90	6.30	0.62	5.19	7.71	0.923	8.61
140	6.35	0.63	6.00	6.08	0.574	4.50	6.79	0.71	5.98	8.31	1.07	9.92
150	6.81	0.72	6.87	6.51	0.659	5.13	7.28	0.82	6.82	8.90	1.23	11.3
160	7.26	0.82	7.79	6.94	0.749	5.80	7.76	0.93	7.72	9.49	1.40	12.8

*(Courtesy, Cameron Hydraulic Data Book.)

On the suction side of the pump, the head loss due to friction (hp) in an 8-ft length of pipe is calculated as follows:

$$hp = (8 \text{ ft})\left(\frac{3.9 \text{ ft}}{100 \text{ ft}}\right) = 0.312 \text{ ft}$$

To calculate the friction loss in fittings, use the same formula used in Solution 9-2; that is:

$$hf = (K)\left(\frac{V^2}{2g}\right)$$

where   hf = Friction in fittings and valve
            V = Water velocity in pipe
                (5.64 ft/sec per Table 9-1)
            g = Acceleration of gravity (32.174 ft/sec^2)
            K = Resistance coefficient for each fitting

therefore:

$$hf = (K)\left(\frac{5.64^2}{2 \times 32.174}\right) = (K)(0.494)$$

On the suction side of the pump, the fittings are a foot valve (a type of check valve) and a 90° elbow at the entrance into the pump (see section A-A in Figure 9-6). Based on the tables listing the values of resistance coefficients included in Appendix B, the applicable value of these K coefficients is 1.4 for the foot valve and 0.54 for the 90' elbow. Insert these values to complete the equation above:

$$hf = (1.4 + 0.54)(0.494) = 0.958 \text{ ft}$$

To calculate total head loss on the suction side, add the static head to the friction losses (pipes and fittings). Remember that the static head in this case is the difference in elevation between the center line elevation of the pump and the water level[2] in the holding tank. (Since we are calculating the losses by dividing the system into the suction and the discharge sides of the pump, the pump center line becomes the reference point between the two sides in this case.)

From Figure 9-6, it is possible to calculate the difference in elevation:

$$31 \text{ ft} - 25 \text{ ft} = 6 \text{ ft}$$

The total suction head, or pressure loss, which is measured in ft, thus becomes:

$$6 \text{ ft (static head)} + 0.312 \text{ ft (friction in pipe)}$$
$$+ 0.958 \text{ ft (friction in fittings)} = 7.27 \text{ ft}$$

Now it is necessary to calculate the friction or head loss in the discharge side of the pump. For the 1350-ft length of 3-in. diameter discharge pipe, the friction in the pipe is:

$$(1350 \text{ ft})\left(\frac{3.90 \text{ ft}}{100 \text{ ft}}\right) = 52.65 \text{ ft}$$

Before calculating the friction in fittings, first tabulate the applicable K-coefficient values (see Table 9-2).

Table 9-2   Friction in Fittings-Resistance Coefficient

Fitting	Value of K	Pipe Diameter, in.
45° elbow	0.29	3
90° standard elbow	0.54	3
Gate valve	0.14	3
Swing-check valve	1.80	3
Pipe exit	1.00	Sharp edge

It is then possible to calculate the friction loss in fittings:

$$hf = (K)\left(\frac{V^2}{2g}\right)$$

$$hf = (0.54 + 0.14 + 1.8 + 0.29 + 0.29 + 0.54 + 0.54 + 1.0)\left(\frac{V^2}{2g}\right)$$

$$hf = (5.14)(0.494) = 2.54 \text{ ft}$$

The static head, or difference in elevation, is 279 ft (310 ft − 31 ft). The static head, in this case, is considered a loss. (As mentioned earlier, the lower reference point in this case is the pump center line.)

Since the flow is against gravity and the static head must be overcome, the total discharge head loss (h_d) is:

$$h_d = 52.65 \text{ ft (loss in pipe)} + 2.54 \text{ ft (loss in fittings)}$$
$$+ 279 \text{ ft (static head)}$$
$$= 334.19 \text{ ft}$$

Add together the losses on both sides of the pump to get the total pressure loss in the system (H_T), as follows:

$$H_T = 7.27 \text{ ft (suction head)} + 334.19 \text{ ft (discharge head)}$$
$$= 341.46 \text{ ft}$$

Table Verification

In Example 9-4, we obtained the pipe friction value from a neat and orderly table. However, these tables are the result of a great deal of calculation based on the Darcy–Weisbach formula. Another similar formula is called the Hazen–Williams empirical formula. The Hazen–Williams formula is referred to as empirical because it is based on laboratory and field observation. The Hazen and Williams pressure-loss formula is:

$$hf = (0.002083)(L)\left(\frac{100}{C}\right)^{1.85}\left(\frac{Q^{1.85}}{d^{4.8655}}\right)$$

where hf = Friction in pipe, ft/100 ft
        0.002083 = Empirically determined coefficient
        L = Length of pipe, ft (in this case, 100 ft)
        C = Roughness coefficient based on the
             pipe material (see Table 9-3)
        Q = Flow, gallons per minute (gpm)
        d = Pipe diameter, in.

Table 9-3 Values of the Constant C Used in Hazen and Williams Formula (New Pipe)

Pipe Material	Values of C		
	Range	Average Value	Normally Used Value
Bitumastic-enamel-lined steel centrifugally applied	130–160	148	140
Asbestos-cement	140–160	150	140
Cement-lined iron or steel centrifugally applied	—	150	140
Copper, brass, or glass as well as tubing	120–150	140	130
Welded and seamless steel	80–150	140	100
Wrought iron, cast iron	80–150	130	100
Tar-coated cast iron	50–145	130	100
Concrete	85–152	120	100
Full-riveted steel (projecting rivets in girth and horizontal seams)	—	115	100
Corrugated steel	—	60	60

Value of C	150	140	130	120	110	100	90	80	70	60
For (100/C) at 1.85 power is	0.47	0.54	0.62	0.71	0.84	1	1.22	1.5	1.93	2.57

Example 9-5 To ensure that the values given in the tables for Example 9-4 are correct, calculate the friction loss using the Hazen and Williams formula and the following data:

$$L = 100 \text{ ft}$$
$$C = 150$$
$$d = 3 \text{ in.}$$
$$Q = 130 \text{ gpm}$$

Solution 9-5 Plug the data into the Hazen and Williams formula, as follows:

$$hf = (0.002083)(100)\left(\frac{100}{150}\right)^{1.85}\left(\frac{130^{1.85}}{3^{4.8655}}\right) = 3.8 \text{ ft}$$

The value used from the table was 3.90 ft per 100 ft, which is very close to the one just calculated (3.80 ft per 100 ft). This exercise demonstrates that, for all practical purposes, the table may be used with confidence.

Alternative Solution There is another, easier and faster way to solve the hydraulic problem given in Example 9-4. It involves using equivalent length for fittings (see Table 9-4). This easier alternative is defined as an equivalent length of straight pipe that has the same friction loss as the respective fitting or valve.

Various piping books and publications may indicate slightly different equivalent length values for the same fitting. These differences are usually small and, therefore, negligible. If a certain type of fitting cannot be found in an available table, an approximate value can be estimated based on a similar fitting. The equivalent value (length) can also be obtained from the fitting's manufacturer.

Example 9-6 Solve the same problem given in Example 9-4, but use the equivalent length for fittings and valves and make the calculation for the total head loss for the entire system (suction and discharge). The system data remain the same.

Solution 9-6 The developed length of pipe, or the actual measured length, is as follows:

1350 ft (discharge) + 8 ft (suction) = 1358 ft

For this application, and based on Table 9-4, the equivalent lengths for fittings and valves are listed below:

Table 9-4 Equivalent Lengths for Pipe Fittings

Nominal Pipe Size, in.	Gate Valve — Full Open	Globe Valve — Full Open	Butterfly Valve —	Angle Valve — Full Open	Swing-Check Valve, Full Open	90° Elbow	Long Radius 90° & 45° Std Elbow	Close Return Bend	Standard Tee — Through Flow	Standard Tee — Branch Flow	Mitre Bend 45°	Mitre Bend 90°
½	0.41	17.6	—	7.78	5.18	1.55	0.83	2.59	1.0	3.1		
¾	0.55	23.3	—	10.3	6.86	2.06	1.10	3.43	1.4	4.1		
1	0.70	29.7	—	13.1	8.74	2.62	1.40	4.37	1.8	5.3		
1¼	0.92	39.1	—	17.3	11.5	3.45	1.84	5.75	2.3	6.9		
1½	1.07	45.6	—	20.1	13.4	4.03	2.15	6.71	2.7	8.1		
2	1.38	58.6	7.75	25.8	17.2	5.17	2.76	8.61	3.5	10.3	2.6	10.3
2½	1.65	70.0	9.26	30.9	20.6	6.17	3.29	10.3	4.1	12.3	3.1	12.3
3	2.04	86.9	11.5	38.4	25.5	7.67	4.09	12.8	5.1	15.3	3.8	15.3
4	2.68	114	15.1	50.3	33.6	10.1	5.37	16.8	6.7	20.1	5.0	20.1
5	3.36	143	18.9	63.1	42.1	12.6	6.73	21.0	8.4	25.2	6.3	25.2
6	4.04	172	22.7	75.8	50.5	15.2	8.09	25.3	10.1	30.3	7.6	30.3
8	5.32	226	29.9	99.8	33.3	20.0	10.6	33.3	13.3	39.9	10.0	39.9
10	6.68	284	29.2	125	41.8	25.1	13.4	41.8	16.7	50.1	12.5	50.1
12	7.96	338	34.8	149	49.7	29.8	15.9	49.7	19.9	59.7	14.9	59.7
14	8.75	372	38.3	164	54.7	32.8	17.5	54.7	21.8	65.6	16.4	65.6
16	10.0	425	31.3	188	62.5	37.5	20.0	62.5	25.0	75.0	18.8	75.0
18	16.9	478	35.2	210	70.3	42.2	22.5	70.3	28.1	84.4	21.1	84.4
20	12.5	533	39.2	235	78.4	47.0	25.1	78.4	31.4	94.1	23.5	94.1
24	15.1	641	47.1	283	94.3	56.6	30.2	94.3	37.7	113	28.3	113
30	18.7					70	37.3	117	46.7	140	35	140
36	22.7					85	45.3	142	56.7	170	43	170
42	26.7					100	53.3	167	66.7	200	50	200
48	30.7					115	61.3	192	76.7	230	58	230

Fittings	Number	Equivalent Length, Each Length, ft	Total Equivalent Length, ft
3-in. foot valve (same as swing-check valve)	1	25.5	25.5
90° elbow	4	7.67	30.68
45° elbow (long radius)	2	4.09	8.18
3-in. gate valve	1	2.04	2.04
3-in. swing-check valve	1	25.5	25.5
Sharp pipe exit (estimated)	1	17.5	17.5
Total			**109.4**

Now add the total equivalent length for fittings to the actual pipe length, as follows:

$$1358 \text{ ft} + 109.4 \text{ ft} = 1467.4 \text{ ft}$$
$$= \text{Total equivalent length of pipe}$$

Therefore, the total friction loss in pipe *and* fittings is:

$$(1467.4 \text{ ft}) \left(\frac{3.9 \text{ ft}}{100 \text{ ft}} \right) = 57.22 \text{ ft}$$

The total difference in elevation of the water levels (static head) is equal to 310 ft – 25 ft, or 285 ft ; therefore, the total head (pressure) loss is 57.22 ft + 285 ft, or 342.22 ft.

Compare the previous value of 341.46 ft, which was obtained in Example 9-4 using the K resistance coefficients, to the value of 342.22 ft, which was obtained using the equivalent length method. The calculations are equivalent; therefore, it is easier to use the equivalent length method when solving plumbing problems. Keep in mind that these results are based on an engineering calculation in which minor approximations are acceptable.

These examples show that, by knowing the flow, pipe diameter, system configuration, and pipe material, it is possible to calculate the head or friction loss in the system. When calculating a problem like this, consider that the piping system will age, so add in a safety factor of 10 to 15% to establish the correct value when selecting the pump. Thus, a value of approximately 380 ft is required for the selection of the pump head (342.22 ft + 37.78 ft). The pump-head value and the pump flow (previously given as 130 gpm) are used in the selection of the pump.

Example 9-7   Calculate the outlet pressure in the system shown in Figure 9-7, in which water flows by gravity. The technical data is as follows:

Water flow (Q): 50 gpm

Pipe diameter: 2 in.

Pipe material: Type K copper tubing

Pipe length: 380 linear ft (developed length, which is the actual measured length of the pipe)

Fittings: Two gate valves, one sudden contraction (from the tank into the pipe), one sudden enlargement (discharge open to the atmosphere), one 30° elbow; one 45° elbow

Difference in elevation: 150 ft (static)

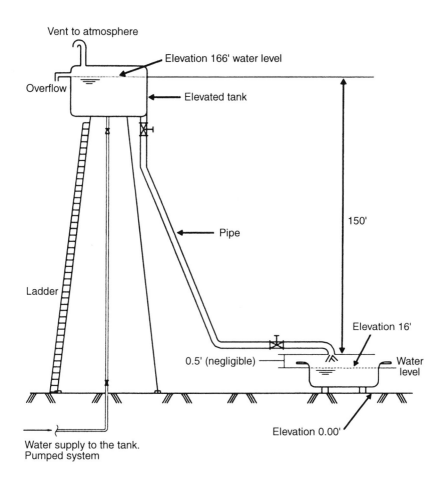

Figure 9-7 Gravity System

Solution 9-7 From the table in Appendix B, the flow of 50 gpm of water in a 2-in. diameter, Type K copper tubing (pipe) has a velocity of 5.32 ft/sec and a friction loss of 5.34 ft per 100 ft.

Use the following table to calculate the equivalent length of pipe method described earlier:[3]

Fittings	Quantity	Each Equiv. Length, ft	Length Total, ft
2-in. gate valve	2	1.38	2.76
Sudden contraction	1	10.3	10.3
Sudden enlargement (estimated length as a standard tee through the branch)	1	10.3	10.3
45° elbow	1	2.76	2.76
30° elbow (estimated to be the same as 45° elbow)	1	2.76	2.76
Total Equivalent Length for Fittings			**28.88** (round to 29)

The total equivalent length of the pipe and fittings is:

$$380 \text{ ft} + 29 \text{ ft} = 409 \text{ ft}$$

Friction loss in the pipe and fittings is:

$$(409 \text{ ft})\left(\frac{5.34 \text{ ft}}{100 \text{ ft}}\right) = 21.84 \text{ ft}$$

Given the difference in the elevation of 150 ft, the outlet pressure is:

$$150 \text{ ft} - 21.84 \text{ ft} = 128.16 \text{ ft}$$

$$\frac{128.16}{2.31} = 55.48 \text{ psi}$$

The difference in elevation (static head) in this gravity-flow type example (or in any other downhill flow) assists the flow, because the weight of the water column pushes the water down toward the discharge. This is the reason the friction is *deducted* from the static head.

Example 9-8 Calculate the same problem given in Example 9-7, only this time with a flow of 200 gpm. All other data remains the same. Based on the table in Appendix B, the velocity is 21.3 ft/sec, and the friction is 65.46 ft/100 ft. *Note: a velocity of 21.3 ft/sec is unacceptable, but it is used here for illustrative purposes.*

Solution 9-8   From Example 9-7, we determined that the equivalent length of pipe and fittings is 409 ft. Therefore, the friction loss in pipe and fittings is:

$$(409 \text{ ft})\left(\frac{65.46 \text{ ft}}{100 \text{ ft}}\right) = 267.73 \text{ ft}$$

Given the difference in the elevation of 150 ft, the outlet pressure has a negative value:

$$150 \text{ ft} - 267.73 \text{ ft} = -117.73 \text{ ft}$$

The result of the calculations means that 200 gpm cannot flow through the system because the pipe diameter is too small and the friction is too high for such flow. Water will flow but only at a maximum rate of 145 gpm. This value can be mathematically calculated as follows:

$$\left(\frac{x \text{ ft}}{100 \text{ ft}}\right)(409 \text{ ft}) = 150 \text{ ft static head}$$

$$x = \frac{(150 \text{ ft})(100 \text{ ft})}{409 \text{ ft}} = 36.67 \text{ ft}$$

From the table in Appendix B, the corresponding flow of a 2-in., type K copper tube at the friction calculated above is approximately 145 gpm.

Notes

[1]The conversion value of 0.433 is derived from the following:

$$0.433 = \frac{14.7 \text{ psi}}{33.96 \text{ ft}} \qquad \text{Reciprocal: } 1 \text{ psi} = 2.31 \text{ ft of H}_2\text{O}$$

[2]Normally, the difference in elevation (static head) is from water level to water level, or to a water discharge outlet elevation.

[3]Some manuals give slightly different equivalent lengths for copper tubing fittings than for steel, but the difference is small enough that the same table may be used.

Chapter 10: Sprinkler-System Calculations

There are two methods used to calculate the size of each pipe supplying water to sprinkler heads:
- *Pipe-schedule method*—Limits the number of sprinklers per branch. (This has limited acceptance by authorities having jurisdiction.)
- *Hydraulic calculations*—Do not limit the number of sprinklers per branch, allowing for the use of smaller pipe sizes.

In the pipe-schedule method, tables found in NFPA Standard no. 13, as applicable to the various hazard types, are used. These tables list the number of sprinkler heads that may be located on each branch and the pipe size to be installed for the respective number of heads. This is important, because sprinkler calculations are made based on the number of heads and their arrangement. The designer then develops the pipe arrangement.

In the pipe-schedule method, information found in Tables 10-1 through 10-3, which are excerpted from NFPA Standard no. 13, is used. These tables list steel and copper-pipe sizes and the number of heads for light, ordinary, and extra hazards.

In a light or ordinary-hazard occupancy, branch lines must not exceed 8 sprinklers on either side of a cross main. There may be some instances when more than 8 sprinklers on a branch line is necessary. In this case, lines may be increased to 9 sprinklers by making the two end lengths 1 in. and 1¼ in., respectively, and the sizes thereafter standard .

Ten sprinklers may be placed on a branch line by making the two end lengths 1 in. and 1¼ in., respectively, and feeding the tenth sprinkler by a 2½-in. pipe.

Table 10-1 Pipe Sizes for Light-Hazard Occupancies

Steel		Copper	
Pipe Size, in.	Number of Sprinklers	Tube Size, in.	Number of Sprinklers
1	2	1	2
1¼	3	1¼	3
1½	5	1½	5
2	10	2	12
2½	30	2½	40
3	60	3	65
3½	100	3½	115
4	See NFPA Standard no. 13*	4	See NFPA Standard no. 13*

(Courtesy NFPA.)

* Areas requiring more sprinklers than the number specified for 3½-in. pipe, without subdividing partitions (not necessarily fire walls), shall be supplied by mains or risers sized for ordinary-hazard occupancies.

Table 10-2 Pipe Sizes for Ordinary-Hazard Occupancies

Steel		Copper	
Pipe Size, in.	Number of Sprinklers	Tube Size, in.	Number of Sprinklers
1	2	1	2
1¼	3	1¼	3
1½	5	1½	5
2	10	2	12
2½	20	2½	25
3	40	3	45
3½	65	3½	75
4	100	4	115
5	160	5	180
6	275	6	300
8	See NFPA Standard no. 13*	8	See NFPA Standard no. 13*

(Courtesy NFPA.)

* When the distance between sprinklers on the branch line exceeds 12 ft or the distance between the branch lines exceeds 12 ft, the number of sprinklers for a given pipe size shall be in accordance with Table 10-3.

**Table 10-3 Pipe Sizes for Ordinary-Hazard Occupancies for
Distances Greater than 12 Ft**

Steel		Copper	
Pipe Size, in.	Number of Sprinklers	Tube Size, in.	Number of Sprinklers
2½	15	2½	20
3	30	3	35
3½*	60	3½*	65

(Courtesy NFPA.)

* For all other pipe and tube sizes, see NFPA Standard no. 13.

The piping schedule in Table 10-4 is reprinted as a guide for existing systems only. New extra-hazard occupancy systems should be designed using hydraulic calculations. The use of pipe schedules for extra-hazard occupancies was prohibited in 1991. However, there are many existing pipe-schedule systems currently in place, so the selection table for extra hazard has been retained.

Table 10-4 Pipe Sizes for Extra-Hazard Occupancies

Steel		Copper	
Pipe Size, in.	Number of Sprinklers	Tube Size, in.	Number of Sprinklers
1	1	1	1
1¼	2	1¼	2
1½	5	1½	5
2	8	2	8
2½	15	2½	20
3	27	3	30
3½	40	3½	45
4	55	4	65
5	90	5	100
6	150	6	170

(Courtesy NFPA.)

Table 10-5 shows the recommended lengths of operation for sprinkler systems based on pipe-schedule calculations and the types of hazard.

Table 10-5 Water-Supply Requirements for Pipe-Schedule Sprinkler Systems

Classification	Minimum Residual Pressure Required, psi	Acceptable Flow at Base of Riser, gpm	Duration, min
Light hazard	15	500 to 750	30 to 60
Ordinary hazard	20	850 to 1500	60 to 90

Note: Open sprinkler and deluge systems must be hydraulically calculated, according to applicable standards.

Pipe-Schedule Method

As previously mentioned, the pipe-schedule method is used to design sprinkler systems. Its use was limited in 1991 to small installations or additions (of any size), existing pipe-schedule systems, and light or ordinary-hazard occupancies. This method may be used in larger light-hazard and ordinary-hazard installations when the water supply is available at a residual pressure of at least 50 psi. The pipe-schedule method was restricted to avoid its use under conditions of marginal water-supply pressure and flow.

Example 10-1 Using the pipe-schedule method, design a sprinkler system for a shoe store that is classified as an ordinary-hazard occupancy. The store has 4000 ft². (The toilet room and manager's office each have an additional sprinkler head.)

Solution 10-1 First determine the number of heads that are required. Based on the type of hazard, 1 head can cover 100 to 130 ft². (One head can cover an area as follows: light hazard 225 ft², ordinary hazard 100 to 130 ft², extra hazard 90 to 100 ft².) To be on the safe side, use 100 ft²:

$$\frac{4000 \text{ ft}^2}{100 \text{ ft}^2/\text{head}} = 40 \text{ heads}$$

Add 1 head for the toilet room and 1 head for the manager's office: 40 + 2 = 42 heads.

According to Table 10-2, the supply pipe needed to satisfy the system is 3½ in. Select the next larger pipe size (4 in.) to accommodate fire-department connections, which require a 4-in. supply pipe (minimum). Perform flow tests to ensure that there are enough flow and pressure to satisfy the system.

An architectural plan is required to establish sprinkler location in the area. The plan and elevation for this particular system could look like those shown in Figures 10-1 and 10-2. Note the following features in these figures:

- Pendent-type sprinkler heads are installed in areas with ceilings; upright sprinkler heads are installed when there is no ceiling.
- Minimum pipe size is 1 in. for steel material and ¾ in. if copper tubing is used.
- Sprinkler heads are arranged in a regular geometrical order.
- Enclosed areas must be provided with separate additional heads for local protection.

Hydraulic calculations could be used to size the pipes for this example, and they might result in smaller pipe sizes. It is up to the owner and the engineer to make a decision as to whether more expensive, hydraulic calculations are worth using.

Hydraulic Calculations

A hydraulic calculation determines flow (gpm) and pressure (head) losses through each pipe section of the sprinkler system based on the water-velocity limitation selected. Systems designed using hydraulic calculations normally cost less (in materials) than those designed using the pipe-schedule method.

Performing hydraulic calculations manually is tedious work, and the possibility of error is relatively high. Fortunately, computer software has been developed to perform hydraulic calculations quickly and easily. In addition, computer software allows the knowledgeable technician to check different water-supply routes and configurations so that the shortest pipe run may be selected, which makes the job more cost efficient.

While hydraulic-software packages are very helpful, the computer operator must still have a good technical background in the area of fire protection. The operator must choose from a selection of parameters, which requires specialized training. One of these

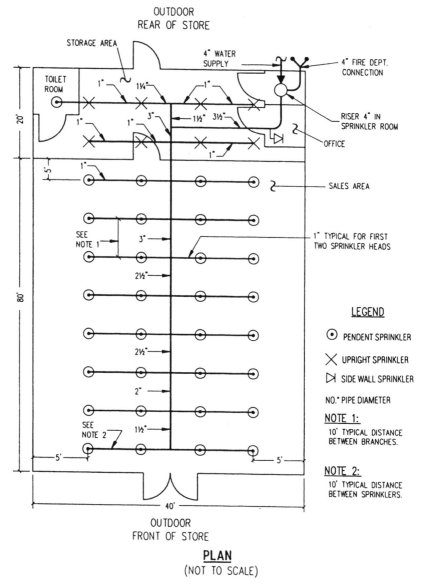

PLAN
(NOT TO SCALE)

Figure 10-1 Sprinkler-System Plan

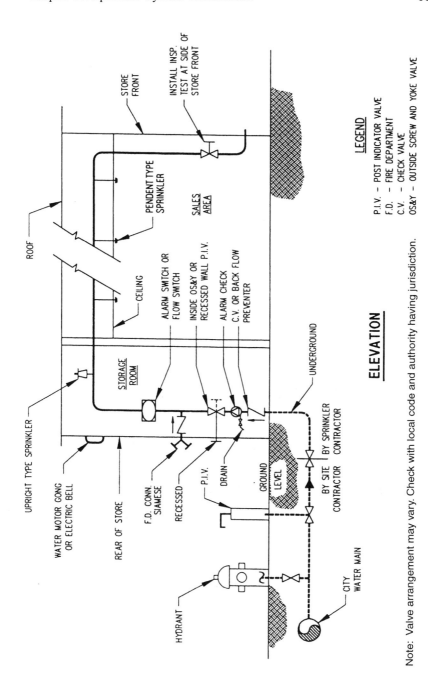

Figure 10-2 Yard and Sprinkler-System Elevation

parameters includes choosing pipe sizes based on pressure-loss (friction) calculations. In a wet-type sprinkler system, the operator must select the water density to be discharged over the fire area, measured in gpm/ft^2 from NFPA Standard no. 13 curves.

To perform a hydraulic calculation on a computer, take the following steps:

- Determine the type of hazard based on occupancy.
- Establish water density (gpm/ft^2) based on NFPA Standard no. 13. (Reproduced density curves are shown in Figure 10-3. Sprinkler-operation area [ft^2] on the water density curve is shown on the vertical, and water density [gpm/ft^2] on the horizontal. Areas located on either hazard-type curve correspond to a particular water density. *Note:* Figure 10-3 may be used only for wet-pipe sprinkler systems and for standard [½-in.] orifice size.)

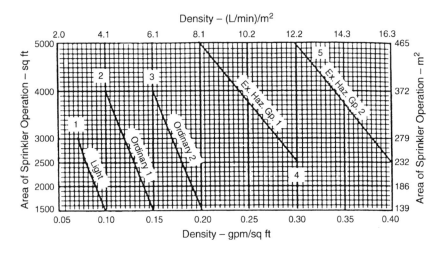

Figure 10-3 Density Curves

For example, consider that a particular warehouse is a light-hazard occupancy and has an area of 22,000 ft^2. For the hydraulic calculation from Figure 10-3, select the area located opposite and farthest away from the point where piping enters the building. In this case, it is 1500 ft^2 at 0.10 gpm/ft^2 water density. However, these are minimum area covered and minimum water density. For an actual calculation, a designer may choose 2500 ft² or any other selected area (larger than the minimum) along the curve for the design area. Similarly, a higher water density than the one read from the graph may be used. Remember, standards and codes list minimum requirements only. A more generous selection gives better coverage, but this comes at a higher cost.

In the 22,000-ft² warehouse, calculations are made for water requirements for the most remote area only. If that area has enough water at the right pressure, the system is correctly designed and every section of the system is covered. It is not expected that an accidental fire will start simultaneously in a few locations, which could be the case if arson is involved.

It is not always necessary for all sprinkler heads at the most remote area to open when a fire occurs. Only the sprinkler heads located directly above the fire will react at high temperatures and subsequently discharge water. However, if an application requires a deluge system, in which all heads are open, the entire area will be completely drenched. This is why deluge systems are on a separate riser, which serves a limited area.

Sprinkler-head spacing and branch spacing must be designed so that each head will cover the predetermined number of square feet in accordance with the code indication for the type of hazard. When a hydraulic calculation is made, the pipe-junction points must be balanced within 0.5 psi, and the highest pressure drop must be carried on (the calculation proceeds from the last, most remote head toward the water-supply pipe entering the building). The equivalent length of pipe should be used for fittings and valves.

Gathering Information

As stated previously, performing manual hydraulic calculations is tedious. The steps required in the calculation given here are lengthy, but they are necessary to ensure correct values. Many of the steps include gathering preliminary material that is necessary even before a calculation can be performed.

The hydraulic calculation example that follows is solved in an orderly manner using a hand calculation. This calculation may be done with a special program on a computer, but it is shown step-by-step here for clarification. No matter which method is used (pipe-schedule or hydraulic), basic information is always required. This is gathered on an information sheet (see Figure 10-4).

Standard abbreviations and symbols are also included with the basic information. These may change from project to project, but Table 10-6 shows the abbreviations used most often.

Table 10-6 Common Abbreviations

Abbreviation	Item
BV	Butterfly valve
Cr	Cross
CV	Swing-check valve
Del V	Deluge valve, alarm valve
E	90° elbow
EE	45° elbow
gpm	U.S. gallons per minute
GV	Gate valve
Lt. E	Long-turn elbow
P_e	Pressure due to elevation difference between indicated points. This can be a positive or negative value.
P_f	Pressure loss due to friction between points indicated in location column
P_n	Normal pressure, in psi, at a point in a pipe
psi	Pounds per square inch
P_t	Total pressure, in psi, at a point in a pipe
P_v	Velocity pressure, in psi, at a point in a pipe
q	Flow increment, in gpm, to be added at a specific location
Q	Summation of flow, in gpm, at a specific location
St	Strainer
T	Tee – flow turned 90°
v	Velocity of water in pipe, in feet per second (fps)
WCV	Butterfly (wafer) check valve

Information Sheet

NAME _____ DATE _____
LOCATION _____
BUILDING _____ SYSTEM NO. _____
CONTRACTOR _____ CONTRACT NO. _____
CALCULATED BY_____ DRAWING NO. _____
CONSTRUCTION: ☐ COMBUSTIBLE   ☐ NON-COMBUSTIBLE   CEILING HEIGHT_____FT.
OCCUPANCY _____

SYSTEM DESIGN

☐ NFPA 13:   ☐ LT. HAZ.   ORD. HAZ. GP. ☐ 1 ☐ 2   EX. HAZ. GP.1☐ 2☐
☐ NFPA 231   ☐ NFPA 231C: FIGURE _____ ; CURVE _____
☐ OTHER (Specify) _____
☐ SPECIFIC RULING _____ MADE BY _____ DATE _____

AREA OF SPRINKLER OPERATION _____	**SYSTEM TYPE**
DENSITY _____	☐ WET ☐ DRY ☐ DELUGE ☐ PRE-ACTION
AREA PER SPRINKLER _____	**SPRINKLER OR NOZZLE**
HOSE ALLOWANCE GPM: INSIDE _____	MAKE _____ MODEL _____
HOSE ALLOWANCE GPM: OUTSIDE _____	SIZE _____ K-FACTOR _____
RACK SPRINKLER ALLOWANCE _____	TEMPERATURE RATING _____

CALCULATION | GPM REQUIRED _____ PSI REQUIRED _____ AT BASE OF RISER
SUMMARY | "C" FACTOR USED: OVERHEAD _____ UNDERGROUND _____

WATER SUPPLY

WATER FLOW TEST	PUMP DATA	TANK OR RESERVOIR
DATE & TIME _____	RATED CAPACITY _____	CAPACITY_____
STATIC PSI _____	AT PSI _____	ELEVATION _____
RESIDUAL PSI _____	ELEVATION _____	
GPM FLOWING _____		**WELL**
ELEVATION _____		PROOF FLOW_____ GPM

LOCATION _____
SOURCE OF INFORMATION _____

COMMODITY STORAGE

COMMODITY_____ CLASS _____ LOCATION _____
STORAGE HEIGHT_____ AREA _____ AISLE WIDTH _____
STORAGE METHOD: SOLID PILED _____% PALLETIZED_____ % RACK _____ %

RACK

☐ SINGLE ROW   ☐ CONVENTIONAL PALLET   ☐ AUTOMATIC STORAGE ☐ ENCAPSULATED
☐ DOUBLE ROW   ☐ SLAVE PALLET   ☐ SOLID SHELVING   ☐ NON-
☐ MULTIPLE ROW   ☐ OPEN   ENCAPSULATED

FLUE SPACING IN INCHES	CLEARANCE FROM TOP OF STORAGE TO CEILING
LONGITUDINAL _____TRANSVERSE _____	_____ FT. _____ IN.

HORIZONTAL BARRIERS PROVIDED _____

Figure 10-4 Information Sheet

An important element in the hydraulic sprinkler calculation is the K factor, called the "water-discharge coefficient." This is a similar coefficient to the one used in the general hydraulic problem worked out in Chapter 9. This discharge coefficient follows the hydraulic laws and varies with the sprinkler-head orifice size as shown in Table 10-7.

Table 10-7 Nominal Sprinkler-Orifice Designation

Size, in.	Designation	K Value
$1/4$	Small	1.3 to 1.5
$5/16$	Small	1.8 to 2.0
$3/8$	Small	2.6 to 2.9
$7/16$	Small	4.0 to 4.4
$1/2$	Standard	5.3 to 5.8
$17/32$	Large	7.4 to 8.2
$5/8$	Extra large	11.0 to 11.5

There is a larger sprinkler orifice with a larger K value, but it is used only for special applications. The orifice size is marked on the sprinkler's frame, and sprinkler manufacturers must furnish the exact K factor for the sprinkler heads furnished for an installation.

Before proceeding with a hydraulic calculation, it is first necessary to obtain basic design information as listed in NFPA Standard no. 13 and/or other applicable standards. When no standards exist, the authority having jurisdiction must be consulted. The basic design information can be obtained from any source available and must include the following:
- Water-supply information.
  — Water-flow data from existing or proposed water source.
  — Location and elevation of static and residual test gauge with relation to the riser reference point.
  — Water-supply location.
- Hydrant tests.
  — Static pressure (psi).

— Residual pressure (psi).
— Flow (available gpm).
— Date of test.
— Time of test.
— Who conducts test and supplies information.
- Water density (gpm/ft^2).
- Area of water application (ft^2). } Based on type of hazard established.
- Area covered by each sprinkler (ft^2).
- In-rack sprinkler demand (if applicable).
- Inside-hose demand (if applicable).
- Drawings that indicate the following:
  — Location.
  — Name of owner and occupant.
  — Description of hazard.
  — Name and address of contractor and/or designer.
  — Name of agency having jurisdiction.
- Detailed worksheet for manual or computer calculation (similar to that shown in Figure 10-4) that indicates the following:
  — Sheet number.
  — Sprinkler description and discharge constant (coefficient K).
  — Hydraulic reference points.
  — Flow in gpm—available.
  — Pipe lengths, center to center of fittings.
  — Friction loss, in psi per ft of pipe.
  — In-rack sprinkler demand.
  — Elevation head, in psi, between reference points.
  — Required pressure, in psi, at each reference point.
  — Notes to indicate starting points or reference to other sheets, or to clarify data shown.
  — Sketch to accompany gridded system calculations to indicate flow quantities determined.

Once the basic information is gathered, the pipe friction losses must be determined based on the Hazen and Williams formula, which was discussed in Chapter 9. To review, the Hazen and Williams pressure-loss formula is as follows:

$$hf = (0.002083)(L)\left(\frac{100}{C}\right)^{1.85}\left(\frac{Q^{1.85}}{d^{4.8655}}\right)$$

where   hf = Friction in pipe, in ft/100 ft
             0.002083 = Empirically determined coefficient
             L = Length of pipe, in ft (in this case 100 ft)
             C = Roughness coefficient based on the
                 pipe material (see Table 9-3)
             Q = Flow, in gpm
             d = Pipe diameter, in.

Example 10-2[1] Design a sprinkler system for a building that measures 130 ft × 200 ft. The building is classified ordinary hazard, Group 1. Use the hydraulic calculation. The floor plan and elevation of the building are shown in Figures 10-5 and 10-6. Other pertinent information is as follows:

> Water density = 0.15 gpm/ft^2
> Covered area = 1500 ft^2
> Coverage = 130 ft^2/head (see footnote #2)
> K factor (sprinkler) = 5.65 (see footnote #3)
> Pipe = Schedule 40 steel
> C coefficient = 120

Solution 10-2 To determine the number of sprinklers required, divide the covered area (1500 ft^2) by the coverage per sprinkler.

$$\frac{1500 \text{ ft}^2}{130 \text{ ft}^2} = 11.54 \text{ sprinklers (round to 12)}$$

To calculate the number of sprinklers per branch line, use the following formula:

$$\frac{1.2\sqrt{A}}{D}$$

where   A = Covered area, ft^2

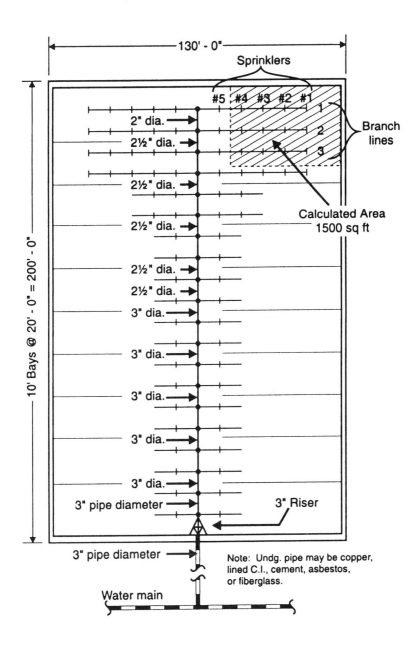

Figure 10-5 Example 10-2 Plan

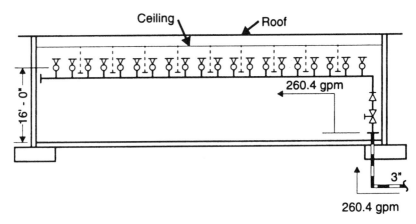

Figure 10-6 Example 10-2 Elevation

$$D = \text{Distance between sprinklers on the branch, ft}$$

Plug in the appropriate numbers:

$$\frac{1.2\sqrt{1500}}{13} = 3.58 \text{ (round to 4)}$$
$$\text{sprinklers per branch line}$$

The 1.2 coefficient in the formula will always make the most remotely covered area a rectangle (as opposed to a square) with the long side parallel to the branch line (see Figure 10-7). In this example, the coverage area of 1500 ft² is the most remote rectangular area. Based on the calculations performed, this area encompasses 4 sprinklers on the branch line and 3 branch lines, to equal the total calculated number of 12 sprinklers.

In Figure 10-7, which is simply a detail from Figure 10-5, the system is symmetrical around the cross main; therefore, this area may be as shown or located in a symmetrical corner of the building. When the most remote or demanding area is not so obviously located, the designer usually prepares and presents three calculated alternatives to support the one selected. The calculation must start at the opposite end of the water source.

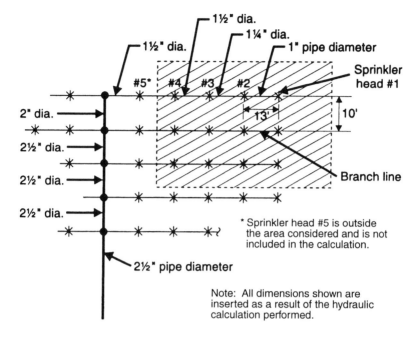

Figure 10-7 Remote-Area Rectangle and Coverage

After performing some additional calculations, as follows, it is possible to start filling in all the information in Figure 10-8.

Flow (Q) for the first sprinkler is determined by multiplying the water density by the coverage per sprinkler:

$$0.15 \text{ gpm/ft}^2 \times 130 \text{ ft}^2 = 19.5 \text{ gpm}$$

Per code requirement, the pipe size between sprinklers no. 1 and no. 2 is 1-in. diameter (see Figure 10-7). The fitting directly connected to the sprinkler is not included in the calculation, so nothing is shown under the "pipe fittings and devices" column of the worksheet. The equivalent pipe length, which is the center-to-center distance between sprinklers, is 13 ft. The pressure loss (psi/ft) is determined by substituting the values of C = 120, L = 1 ft, diameter = 1 in., and flow (Q) = 19.5 gpm in the Hazen and Williams formula. The result is:

$$P = \frac{(5.10)(19.5)^{1.85}}{10^4} = 0.124 \text{ psi/ft}$$

Therefore, the pressure loss from sprinkler no. 1 to sprinkler no. 2 is as follows:

$$(0.124 \text{ psi/ft}) (13 \text{ ft}) = 1.6 \text{ psi}$$

The total pressure (P_t) is determined by using the following formula:

$$P_t = \left(\frac{Q}{K}\right)^2$$

Therefore,

$$\left(\frac{19.5}{5.65}\right)^2 = 11.9 \text{ psi (total pressure}$$
required at sprinkler no. 1)

The pressure at sprinkler no. 2 is:

$$11.9 \text{ psi} + 1.6 \text{ psi} = 13.5 \text{ psi}$$

The flow (Q) from sprinkler no. 2 is equal to:

$$Q = K\sqrt{P}$$

By substituting the known values, we obtain:

$$5.65\sqrt{13.5} = 20.7 \text{ gpm}$$

The flow at sprinkler no. 2 is added to the flow from sprinkler no. 1, and the calculation for flow and pressure continues, per the procedures used above. Therefore, the combined flow of sprinklers no. 1 and no. 2 is as follows:

$$19.5 \text{ gpm} + 20.7 \text{ gpm} = 40.2 \text{ gpm}$$

In order to maintain the system at approximately the same pressure drop, the pipe size has to be increased as the flow increases. For a flow of 40.2 gpm, 1¼-in. pipe will be used (see Figure 10-7). The pipe length remains 13 ft, and the calculated pressure loss is 0.124 psi/ft. Therefore, the pressure at sprinkler no. 3 is as follows:

$$13.5 \text{ psi} + 1.6 \text{ psi} = 15.1 \text{ psi}$$

The flow (Q) at sprinkler no. 3, based on the above formula, is equal to:

$$5.65\sqrt{15.1} = 21.95 \text{ gpm (round to 22 gpm)}$$

The combined flow of sprinkler nos. 1, 2, and 3 is as follows:

$$40.2 \text{ gpm} + 22 \text{ gpm} = 62.2 \text{ gpm}$$

To maintain for this flow a pressure drop close to the one calculated for the first two sprinkler heads using the Hazen and Williams formula (or a table), the pipe size is 1½ in. In a 1½-in. pipe flowing 62.2 gpm, the pressure loss is 0.132 psi/ft.

The pressure loss between sprinkler nos. 3 and 4 is as follows:

$$(13 \text{ ft}) (0.132 \text{ psi/ft}) = 1.7 \text{ psi}$$

The total pressure loss is as follows:

$$15.1 \text{ psi} + 1.7 \text{ psi} = 16.8 \text{ psi}$$

Flow (Q) at sprinkler no. 4 is as follows:

$$5.65\sqrt{16.8} = 23.2 \text{ gpm}$$

The total flow (Q) for the 4 sprinklers on the most remote branch is 85.4 gpm. The flow in sprinkler no. 5 is ignored because it is outside the area considered in the calculation. If the 12 heads calculated are covered, any other head or equal area in this system is covered.

Now that the sprinkler information has been calculated, it is necessary to determine the K factor for branch-water flow. This calculated K factor is just a flow coefficient independent of the sprinkler-head orifice. Use the same formula as before:

$$Q = K\sqrt{P} \text{ or } K = \frac{Q}{\sqrt{P}}$$

While Q is known (85.4 gpm), $\sqrt{P}$ must be calculated before the problem can be solved. The pressure drop is calculated by adding the pressure at sprinkler head 4 (16.8 psi) plus the pressure drop in the pipe to the cross main, the pressure drop in the fittings, and the pressure drop in the rest of the 1½-in. pipe flowing 85.4 gpm:

$$16.8 \text{ psi} + (20.5 \text{ ft} + 16.0 \text{ ft}) (0.237 \text{ psi/ft}) = 25.4 \text{ psi}$$

where   Pipe lengths = 20.5 ft plus 16 ft equivalent
length for fittings
Total equivalent length = 36.5 ft

Therefore, the calculated K factor is:

$$\frac{85.4}{\sqrt{25.4}} = 16.95$$

There is a 2-in. diameter pipe between branches 1 and 2 (Figure 10-5), and the distance between branches is 10 ft. The pressure drop in a 2-in. steel pipe flowing 85.4 gpm (calculated or obtained from a table) is 0.07 psi/ft, so the pressure required at branch 2 is as follows:

$$25.4 \text{ psi} + (0.07 \text{ psi/ft} \times 10 \text{ ft}) = 26.1 \text{ psi}$$

The flow (Q) at branch 2 is:

$$16.95\sqrt{26.1} = 86.6 \text{ gpm}$$

The flow in branch 1 plus branch 2 is:

$$85.4 \text{ gpm} + 86.6 \text{ gpm} = 172.0 \text{ gpm}$$

From branch 2 to branch 3, the pipe diameter is increased to 2½ in., and the pressure loss for the 10-ft length, when it is flowing 172 gpm, is 1.09 psi. Therefore:

$$\text{Total pressure drop} = 26.1 \text{ psi} + 1.09 \text{ psi} = 27.2 \text{ psi}$$

To determine the flow in the third branch, we use the same formula:

Flow = $16.95\sqrt{27.2}$ = 88.40 gpm
Total flow = 172.0 gpm + 88.40 gpm. = 260.4 gpm
Total flow for 12 sprinklers = 260.4 gpm
Pressure required at branch 3 = 27.2 psi

CONTRACT NAME **GROUP I** 1500 φ   SHEET___OF___

STEP NO.	NOZZLE IDENT. AND LOCATION	FLOW IN G.P.M.	PIPE SIZE	PIPE FITTINGS AND DEVICES	EQUIV. PIPE LENGTH	FRICTION LOSS P.S.I./FOOT	PRESSURE SUMMARY	NORMAL PRESSURE	D=0.15 GPM/φ K=5.65 NOTES	REF. STEP
1	1 BL-1	q	1		L 13.0	C=120	Pt 11.9	Pt		
					F		Pe	Pv	q=130X.15=	
		Q 19.5			T 13.0	.124	Pf 1.6	Pn	19.5	
2	2	q 20.7	1¼		L 13.0		Pt 13.5	Pt		
					F		Pe	Pv	q=5.65√13.5	
		Q 40.2			T 13.0	.125	Pf 1.6	Pn		
3	3	q 22	1½		L 13.0		Pt 15.1	Pt		
					F		Pe	Pv	q=5.65√15.1	
		Q 62.2			T 13.0	.132	Pf 1.7	Pn		
4	4 DN RN	q 23.2	1½	2T-16	L 20.5		Pt 16.8	Pt		4
					F 16.0		Pe	Pv	q=5.65√16.8	
		Q 85.4			T 36.5	.237	Pf 8.6	Pn		
5	CM TO BL-2	q	2		L 10.0		Pt 25.4	Pt	K= 85.4/√25.4	5
					F		Pe	Pv		
		Q 85.4			T 10.0	.07	Pf .7	Pn	K=16.95	
6	BL-2 CM TO BL-3	q 86.6	2½		L 10.0		Pt 26.1	Pt		6
					F		Pe	Pv	q=16.95√26.1	
		Q 172.0			T 10.0	.109	Pf 1.1	Pn		
7	BL-3 CM	q 88.4	2½		L 70.0		Pt 27.2	Pt		
					F		Pe	Pv	q=16.95√27.2	
		Q 260.4			T 70.0	.233	Pf 16.3	Pn		
8	CM TO F15	q	3	E5	L 119.0		Pt 43.5	Pt		8
				AV15	F		Pe 6.5	Pv	Pe=15 x .433	
		Q 260.4		GV1	T 140.0	.081	Pf 11.3	Pn		
9		q	3	E5	L 50.0	C=150	Pt 61.3	Pt	COPPER	9
				GV1	F 32.0	TYPE "M"	Pe	Pv	21 x 1.51=32	
		Q 260.4		T15	T 82.2	.061	Pf 5.0	Pn		
		q			L		Pt 66.3	Pt		
					F		Pe	Pv		
		Q			T		Pf	Pn		
		q			L		Pt	Pt		
					F		Pe	Pv		
		Q			T		Pf	Pn		
							Pt			

LEGEND

BL = BRANCH LINE
RN = RISER NIPPLE

K COEFFECIENT FROM ITEM 5 ON, IS CALCULATED

DN = DOWN
L = PIPE LENGTH
F = EQUIV. LENGTH FOR FITTING

C = ROUGHNESS COEFFECIENT

Figure 10-8 Worksheet

Since the flow of 260.4 gpm is now in the 2½-in. pipe, and the straight pipe is 100 ft or longer, the recommended practice is to increase the pipe diameter one size. Thus, the remaining 105 ft of main pipe will be 3 in. in diameter. This follows a practical rule, which recommends installing a pipe one size larger on a long run of pipe.

The total friction or pressure drop is calculated, including the pipe and fittings for the total flow of 260.4 gpm, and carried to the street connection. This total pressure drop is 66.3 psi (see Figure 10-8). (This pressure will ensure that the last sprinkler head will have 11.9 psi, as previously calculated.)

Summary

A calculation may be done in the manner shown only if the branch lines are basically level, as they are assumed to be in this case. In buildings with pitched roofs or with branch lines at different elevations, the pressure due to elevation (static head) has to be applied as it occurs.

In this problem, the supply pipe outside and beyond the building wall is copper pipe and has a C factor of 150. The city water supply delivered is based on 90 psi static head and 60 psi residual pressure when 1000 gpm is flowing.

It is necessary for the fire-protection technician to understand a manually prepared hydraulic calculation to be able to perform one or check one done by a computer. For information purposes, a computer printout of a hydraulic calculation is included in Appendix E.

Notes

[1]This example is reproduced, with permission, from National Fire Protection Association (NFPA), 1991, *Automatic Sprinkler Systems Handbook*, 5th ed., Quincy, MA: NFPA.

[2]This figure is selected, since the distance between heads is 13 ft and the distance between branches is 10 ft.

[3]The K factor is indicated by the sprinkler manufacturer and is equal to 5.65. The orifice is ½-in. diameter (standard). Based on the table showing the K value for ½-in. orifice, the K value range is 5.3 to 5.8. In this case, the value used, 5.65, is within the range.

Chapter 11: General Information about Fire-Protection Systems

This chapter consists of general information concerning fire-protection system installation and maintenance. The information is given in an abbreviated format for better understanding.

General

A sprinkler system must be designed and installed in accordance with NFPA Standard no. 13, *Installation of Sprinkler Systems*, and it must be properly tested and maintained in accordance with NFPA Standard no. 13A. If properly installed and maintained, the sprinkler system will operate efficiently when needed. The owner of the building is responsible for the working condition of the sprinkler system.

Permissible Leakage during Hydrostatic Tests
- *Above ground*—None.
- *Underground*—Some acceptable, per standards (determination is made by pumping from a calibrated container).

Fire-Department Connection (Sometimes Called "Siamese Connection")
- Contact fire department before specifying the type of connection to ensure threads are compatible. Siamese connection riser diameter must be 3 in. for a 3-in. riser (unusual), otherwise 4 in. with two 2½-in. connections (usual).

- Siamese connection riser diameter must be 3 in. for a 3-in. riser (unusual), otherwise 4 in. with two 2½-in. connections (usual).
- Installation height—see Figure 11-1.
- Fire-department (FD) connection (at least one 1½ in. or 2½ in.) must be installed when an automatic sprinkler system has more than 20 heads. FD connection is not an automatic water source.
- Areas with contents that promote fast-burning fires (e.g., lint, combustible residue, hydraulic fluids, etc.) must have maximum area protection equal to the whole area. In other words, a separate riser must be assigned for that area.

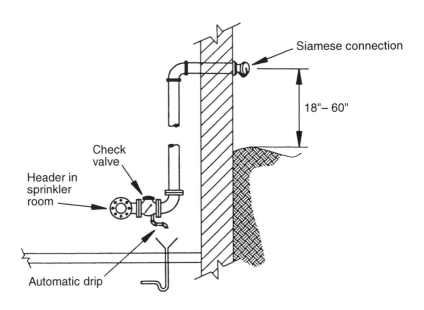

Figure 11-1 Fire-Department Connection

Common Water Supply (Domestic and Fire)

The fire water line must include a backflow preventer to protect potable water from potential contamination from stagnant fire-protection water.

Long Run of Pipe

- If a sprinkler system is fed through a long run of pipe (longer than 100 linear ft), the main must be one pipe size larger than the one resulting from calculations. (See the hydraulic calculation.)
- In general, no water meters are required or recommended for a fire-protection system, because the pressure drop through a water meter is substantial and will add unnecessary power (hp) consumption at the pump motor. However, if required, water meters should be the appropriate type and UL approved.

Sprinkler-Pipe Protection

- *Impact*—Select pipe routing and sprinkler location to protect pipe and sprinklers from impact with moving materials or vehicles.
- *Guards and shields*—Sprinkler heads that may occasionally be hit or mechanically injured should be provided with guards and/or shields. Sprinklers located under grating or on lower storage shelves must have shields. The shield above the sprinkler head slows the flow of heat upward at the head, allowing heat to accumulate around the head so it can respond to the heat from a fire. The shield also protects the sprinkler head from water and/or objects that might fall from upper shelves.
- *Freezing*—Protect sprinkler systems located in unheated areas from freezing by installing a dry-pipe system, a pre-action system, or a system containing antifreeze with insulated pipes and frost-proof casings. Heat tracing of fire-protection pipes is not recommended. The sprinkler/fire-pump room must always be heated.
- *Corrosion*—Prevent pipe corrosion by using corrosion-protective paint. If corrosion protection is necessary, the manufacturer must provide sprinkler heads with a special protective coating. For installations near a saltwater reservoir (sea or ocean), the entire system must be protected against

the corrosive influence of salt aerosol. In areas where corrosion might occur, light wall pipe must not be used; Schedule 40 steel pipe (or heavier) must be used. Painting an installed sprinkler system in the field is prohibited.

- *Earthquake*—Sprinkler-system design must include earthquake design compatible with local regulations.
- *Clearances*—Provide a clearance of 1 in. around the pipe when pipe diameter is 1 in. to 3½ in. and a clearance of 2 in. when pipe diameter is 4 in. or larger.

Joining Fire-Protection Pipes

- Victaulic coupling.
- Welding must comply with Standard AWS 10.9 of the American Welding Society.

Deflector Position

- Always install in a position parallel to the ceiling. If ceiling is slanted, the head and deflector must be slanted (see Figure 11-2).
- If the slanted part is a skylight, higher-temperature sprinklers must be selected to prevent operation due to heat from the sun.

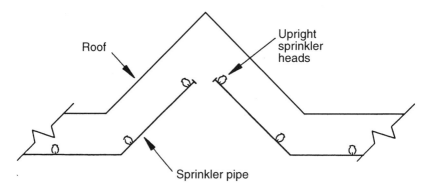

Figure 11-2 Deflector Position

Vertical Shafts
- Install one sprinkler head at the top of the shaft.

Shipping Docks
Separate piping that serves the dock must be protected from freezing. If an antifreeze type system is selected, then consult the antifreeze mixing percentages by volume and specific gravity given by NFPA standards, which depend upon the lowest temperature expected. An alternative is a separate, dry-system riser.

Sprinklers for Protection Against Exposure Fires (Outdoor Protection)
When clusters of buildings, townhouses, stores, etc., are built with combustible materials (wood) and in close proximity to each other, protection against adjacent exposure fires might become necessary. The intensity and other details of fire development at such a site require experienced and educated engineering judgment. Outdoor sprinkler heads must be installed to protect the exposed walls of adjacent buildings. Sprinkler construction is similar to that used for loading-dock protection (dry-type heads). Heat detectors must be employed, preferably the fixed-temperature, rate-anticipation, or rate-of-rise type. Sprinklers must be tested every 2 years during warm weather.

Spare-Sprinkler Requirements
When a system is installed, spare-sprinkler heads must be furnished and stored in a special cabinet together with the installing wrench(es). The number of spare heads is as follows:
- A minimum of 6 for up to 299 heads installed.
- A minimum of 12 for 300 to 999 heads installed.
- A minimum of 24 for 1000 or more heads installed.

Special Types of Sprinkler
- Large-drop sprinklers were developed to fight rapidly burning, high-heat, fuel fires. This type of sprinkler is characterized by a large K factor, of 11.0 to 11.5, and has the proven ability to penetrate a high-velocity fire plume. There

may be combinations of fast-response and large-drop sprinklers. Tests performed with large-drop sprinklers show that 1 head discharges up to 50 gpm, versus the 8 to 20 gpm of a regular sprinkler head. In a large-drop sprinkler, there is a discharge pressure of 50 psi at the farthest head in the system.

Pipes must be sized based on hydraulic calculation in conjunction with the respective design criteria. Large-drop sprinklers can be of the wet, dry, or pre-action type. The installation of a steel-pipe, dry type requires that the piping be internally galvanized. The distance between sprinklers is a maximum of 12 ft and a minimum of 8 ft. Stockpile should not be any closer than 36 in. from the sprinkler.

- Extended-coverage sprinklers were designed to provide coverage for residential areas. They are designed for minimum protection.
- Early suppression and fast response (ESFR) sprinklers can be only of the wet type and must be installed where the roof slope does *not* exceed 1 in./ft. The distance between sprinklers is a maximum of 12 ft and a minimum of 8 ft. The clear space below the sprinkler head is 3 ft.

Valves

- All valves in sprinkler and standpipe systems must be numbered and tagged, and a log must be kept in the building's managing office.
- All control valves must be periodically inspected.
- Sealed valves must be inspected weekly, and locked valves must be inspected monthly.

Sprinkler Testing

High-temperature sprinkler heads must be tested 5 years after installation and every 10 years thereafter by a specialized laboratory acceptable to the authority having jurisdiction.

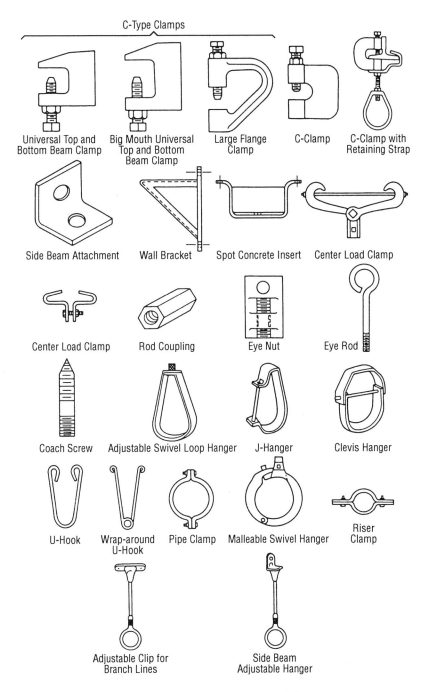

Figure 11-3 Acceptable Hangers and Supports

Pipe Supports

In addition to correct pipe sizes, pipe, and fitting material, sprinkler systems must have properly selected and designed pipe hangers and/or supports. Acceptable hangers are shown in Figure 11-3.

Water-filled pipe weighs substantially more than empty pipe; therefore, supports must be sized for five times the load and include an additional safety factor (e.g., 250 lb at each support). In the sizing of supports, it is also necessary to consider that during operation a pipe may move and/or vibrate slightly due to water flow. Surrounding influences, such as heavy equipment operating in the area, may also cause some vibration of the fire lines, which are normally suspended from the roof or floor above. Well-selected or designed piping supports or restraints must prevent these forces from upsetting a fire-protection piping system. Tle underground pipes (yard) must be laid in a firm bedding so that future, uneven settlement can be avoided. Pipes running under roads should be enclosed in concrete. Pipes penetrating the foundation must be located in pipe sleeves. *Note:* Nothing (pipes, ducts, wires, lamps, etc.) may be hung or supported by the sprinkler piping system.

Fire-Protection Symbols and Abbreviations

Symbols and abbreviations are used on fire-protection drawings and are a graphic representation of a written description. They convey information and simplify the drawings. Fire-protection symbols are shown in Appendix D.

Fire-Protection Definitions

Each NFPA standard includes a list of definitions pertaining to that standard. To clarify various applications, read these definitions.

Fire Rating

Technicians in various disciplines may encounter a fire rating for certain materials. The numbers usually used are 25 for flame spread and 50 for smoke development for a certain class of materials (e.g., upholstery, rugs, and covering materials). The 25 rating means that, when a certain material is set afire, the fire can travel at a rate equal to one-quarter the time and speed of red oak on fire.

The 50 rating means that, during testing, the respective material was burned in a test tunnel where red oak was also burned, and the material produced half the smoke obscuration of red oak during a 10-min period.

Fire Loading

"Fire load" is the total amount of combustibles present in a given area and is usually expressed in pounds per square foot (lb/ft^2). Naturally, the combustible type is also very important, because it influences the amount of heat and gases released. For example, if 10 lb/ft^2 of combustibles has a heat-release capability of 100,000 Btu/hr, the combustible is a light-hazard occupancy.

Piping Materials

Pipe materials to be used for a sprinkler system must meet the characteristics listed in the NFPA 13 table entitled "Pipe and Tube Materials and Dimensions." The smallest steel-pipe schedule acceptable is steel Schedule 10 (used for up to 5 in. diameter pipe). The NFPA also gives standard requirements for fittings.

Sprinkler Arrangement

The piping configuration selected for a project might influence the type of calculation applicable. There are three usual types of sprinkler piping arrangement (see Figure 11-4):

- *Tree arrangement*—Branch lines are connected to one main.
- *Looped arrangement*—Cross mains are interconnected.
- *Gridded arrangement*—Branch lines are interconnected with the cross main(s).

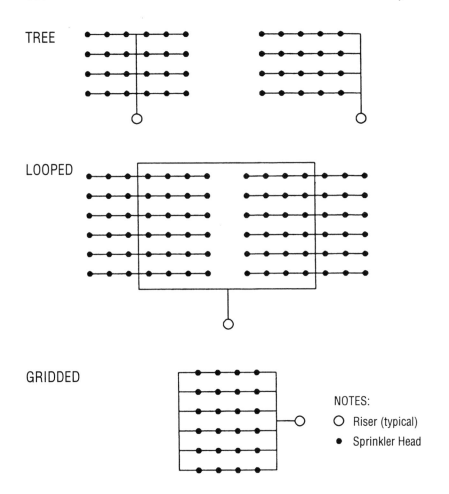

Figure 11-4 Common Sprinkler-Piping Arrangements

Chapter 12: Fire-Protection System Inspection and Maintenance

After an installation is completed by a contractor, the owner's representative must visit the site and perform an installation inspection before the system is covered. A completed fire-protection system is not visible in its entirety because pipes, certain valves, fittings, connections, etc., are sometimes above ceilings, behind walls, etc.

The representative prepares a punch list based on observations made of the system. A punch list consists of the installation deficiencies. Observers look, at a minimum, for the following items:

- Proper system installation.
- Quality of workmanship, piping, fittings, equipment.
- Leaks.
- Properly functioning equipment.
- Pipe slopes, pipe sizes, and accessibility.
  Sprinkler-head spacing conforming to drawings.
- Sprinkler-head type(s) installed in accordance with the specification.
- Hose station and portable extinguisher.
- Accessibility of the fire-department connection.
- Pipe supports and hangers.
- Spare-sprinkler cabinet, number and type of heads, and the required wrench(es).
- Gas-suppression system (if installed) conforming to respective NFPA standard.
- Type, number, and arrangement of fire detectors.

In sum, the representative checks everything to ensure that the system was installed in accordance with project documents, drawings, and specifications. Once the inspector has prepared a punch list, the contractor/installer must correct the deficiencies. (It is difficult to imagine a system that is perfectly installed from the beginning.)

After the fire-suppression system is completely installed, it must be checked again, and after the final corrections are finished, the system is taken over by the owner and is ready for operation when needed. However, the owner and/or user are responsible for the system maintenance program.

Equipment reliability depends on adequate and sustained maintenance. Without proper care, the equipment is bound to fail at some point. In order to perform the required maintenance, technicians must be familiar with the installed system. This requires on-site manufacturer installation, maintenance, and operation instructions.

Maintenance

Maintenance activities can be divided into five categories:
- Inspecting.
- Testing.
- Cleaning.
- Preventive maintenance.
- Repair and replacement.

Inspecting

Inspection schedules are usually generated by the owner (or owner's representative) and are based on manufacturer recommendations for the particular equipment. Inspections must be conducted to identify early warning signs of failure. A weekly inspection should be made of any exposed parts, piping, valves, backflow preventers, hangers and supports, etc. It is important to note any leaks, discoloration, rust, or incorrect position in any of these components. This inspection should be performed by someone who is trained to know what to observe. Of particular importance are valves, most of which must be in a permanently open position.

The weekly inspection of a fire-protection system helps to eliminate problems such as:

- Blocked fire-department connections.
- Vandalized hydrants.
- Leaking pipes and hoses.
- Missing nozzles.
- Permanently open valves that are partially closed.
- Blocked or padlocked emergency exits, etc.
- Freeze-ups (in the winter).

It is also important to check the following points:
- All gauges (monthly).
- Priming of water (when required).
- Clean, dry system valves (not full of grease and dirt).
- System air or nitrogen pressure (weekly).
- All control valves, including sealed valves (weekly) and locked valves (monthly).

After a system is installed and the building is occupied, a fire-safety objective plan must be prepared (see Figure 12-1). Any modifications, inspections, problems, etc., that are connected to the fire-protection system must be recorded and kept in a file in the form of a dated report. These inspection reports and records must be available and kept up-to-date.

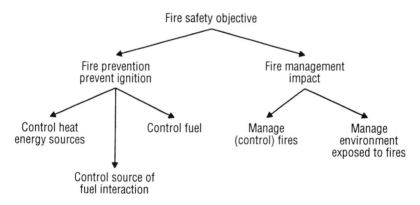

Figure 12-1 Fire-Safety Objective Plan

Testing

The person in charge of the system must test it periodically, based on practical experience and/or manufacturer recommendations, to ensure that the equipment meets specification requirements. All equipment testing must include performance and safety checks.

Alarms must be tested on a regular schedule, which must be well publicized to building occupants. Dry-pipe systems must be tested annually but mainly before the winter time. Table 12-1 illustrates the test and inspection frequency of water-suppression systems.

Cleaning

A scheduled cleaning program is required. Maintenance personnel must perform basic cleaning duties for each system on a regular basis. All parts of the fire-protection system must be kept clean and free of debris.

Preventive Maintenance

All fire-protection equipment must be scheduled for preventive maintenance based on regular inspection results and a scheduled preventive maintenance program.

Repair and Replacement

As a system ages, there is a need for repair and perhaps equipment replacement. It is necessary to maintain spare parts and provide for their storage. In general, the maintenance operation of a system is divided into two categories: breakdown maintenance and preventive maintenance.

The definition of maintenance given in NFPA Standard no. 13A is: "work performed to keep equipment operable or to make repairs." For portable fire-extinguisher maintenance, NFPA Standard no. 10 states: "a thorough examination of the extinguisher is intended to give maximum assurance that the system will operate effectively and safely. It includes a thorough examination and any necessary repair or replacement." This definition can be extended and may be applicable to a fire-protection system instead of just an extinguisher.

Table 12-1 Maintenance and Test Schedule

Parts	Activity	Frequency
Flushing piping	Test	5 years
Fire-department connections	Inspection	Monthly
Control valves		
• Sealed	Inspection	Weekly
• Locked	Inspection	Monthly
• Tamper switch	Inspection	Monthly
Control valves	Maintenance	Annually
Main drain	Flow test	Quarterly
Open sprinklers	Test	Annually
Pressure gauge	Calibration test	
Sprinklers	Test	50 years
High-temperature sprinklers	Test	5 years
Residential sprinklers	Test	20 years
Water-flow alarms	Test	Quarterly
Pre-action/deluge detection system	Test	Semiannually
Pre-action/deluge system	Test	Annually
Antifreeze solution	Test	Annually
Cold-water valves	Open & close valves	Fall—close; Spring—open
Dry/pre-action/deluge systems		
• Air and water Pressure	Inspection	Weekly
• Enclosure	Inspection	Daily—cold weather
• Priming water level	Inspection	Quarterly
• Low-point drains	Test	Fall
• Dry-pipe valves	Trip test	Annually—spring
• Dry-pipe valves	Full-flow trip	3 years—spring
• Quick-opening devices	Test	Semi-annually

(Courtesy, NFPA no. 13A)

Chapter 13: Carbon Dioxide (CO_2) and Halon Replacement

Carbon dioxide is naturally present in the atmosphere in very small amounts (0.03%) and is a normal product of human and animal metabolism. However, an increase in its concentration in the air (to beyond 6% or 7%) is dangerous for humans.

At room temperature, carbon dioxide is a gas that:
- Is colorless and odorless.
- Is inert and electrically nonconductive.
- Is noncorrosive and not damaging.
- Leaves no residue to clean after discharge.
- Does not have to be drained out and/or does not wash chemicals into the drainage system.

Fire-Suppression Agent

Carbon dioxide is a suitable fire-suppression system. It is approximately 50% heavier than the air, which is why nozzles must be located at the upper portion of the protected area; CO_2 moves slowly downward.

CO_2 is liquefied by compression and cooling and converted to a solid state by cooling and expanding. An unusual property of carbon dioxide is the fact that it cannot exist as a liquid at pressures below 60 psig (75 psia). This pressure is known as the "triple-point pressure," at which carbon dioxide may be present as a solid, liquid, or vapor. Below this pressure, it must be either a solid or a gas, depending on the temperature.

If the pressure in a CO_2 storage container is reduced by bleeding off vapor, some of the liquid will vaporize and the remaining liquid will become colder. At 60 psig, the remaining liquid will be converted to dry ice at a temperature of –69°F (–56°C). Further reduction in the pressure will convert all the material to dry ice, which has a temperature of –110°F (–79°C).

The same process takes place when liquid carbon dioxide is discharged into the atmosphere; a large portion of the liquid flashes to vapor, with a considerable increase in volume. The rest is converted into finely divided particles of dry ice at –110°F. It is this dry ice or "snow" that gives the discharge its typically cloudy, white appearance. The low temperature also causes water to condense from the air, so that ordinary water fog tends to persist for a while after the dry ice has evaporated.

When discharged into an enclosed area, carbon dioxide develops a cloud or fog, which is due to the condensation of forming dry ice. The dry ice disappears shortly, which is why the extinguishing by cooling is minimal.

When CO_2 is discharged into an enclosed area at 34% concentration by volume, the temperature in the area drops nearly 80°F very quickly, but the temperature immediately begins to rise. In 2 min, temperatures rise 35°F, in 6 min, 50°F. The temperatures slowly continue to rise to surrounding levels. The extinguishing effect occurs because the oxygen content in the surrounding air is reduced below the 15% threshold, which is the percentage below which combustion cannot take place.

When CO_2 is discharged on electrical equipment, it does not produce an electrical shock. It also does not spread the fire to surrounding areas, which may happen when a fire hose with a solid stream is used. However, if a stream of CO_2 directly hits an operating piece of hot equipment, thermal shock and damage could result.

CO_2 may be used in the following applications:
- Protection of flammable liquids and gases.
- Electrical hazards: computer rooms, transformers, generators, and switch-gear rooms.
- Ovens, broilers, ranges, and kitchen-stove exhaust ducts.

- Application to combustibles with unique value (e.g., legal documents, films, books).

CO_2 should not be used in the following areas:
- When oxidizing materials (chemicals containing their own oxygen supply) are present.
- Where personnel cannot be quickly evacuated.
- When reactive metals are present (e.g., sodium, potassium, magnesium, and titanium).

CO_2 is stored in either high or low-pressure containers. High-pressure containers store CO_2 at 850 psi and 70°F, and each cylinder may weigh 5, 10, 15, 20, 25, 35, 50, 75, 100, or 125 lb. The CO_2 content per cylinder is 60 to 68%, and the balance within the cylinder is an inert propellant gas. Figure 13-1 shows the typical arrangement of high-pressure containers. Low-pressure containers store CO_2 in refrigerated tanks at 300 psi and 0°F.

The conventional break point between high and low-pressure systems is based on the amount of CO_2 required for protection and the space occupied by the cylinders. Normally this is 2000 lb of CO_2. Due to energy conservation, high-pressure systems that do not require refrigeration are used in larger systems. The space occupied by the cylinders is the limiting criteria.

A CO_2 system may be controlled by either an automatic pneumatic or heat-actuator detector (HAD). Detectors may be either electrical or mechanical. For manual operation, a pull cable is used in a mechanical system, a push button is used in an electrical system, and plant or bottled air is used in a pneumatic system. A manual, emergency actuation is used if the automatic operation fails.

When installing a CO_2 system, consider the following points (Appendix C contains a list of elements to be obtained by a contractor installing a CO_2 system):
- High-pressure cylinders must be stored at temperatures of no more than 120°F and no less than 32°F.
- Distribution piping must be steel: For a high-pressure system of ¾ in. and less, use Schedule 40; for 1 in. and larger,

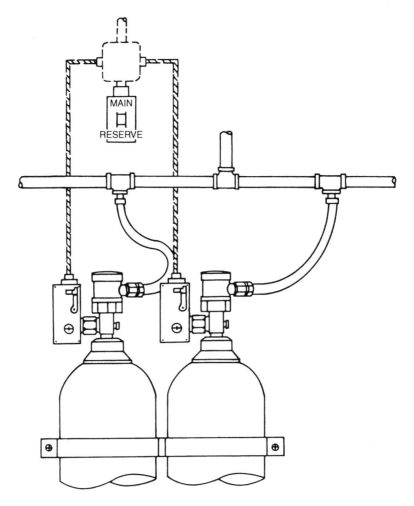

Figure 13-1 CO$_2$ Cylinder Arrangement

use Schedule 80 with malleable and forged-steel fittings. For low-pressure piping, check the schedule with NFPA standards.

• Valves, nozzles, and specialties must be furnished by the vendor and be UL approved or listed.

System Applications

Types of CO$_2$ system application include the following:

- Total flooding in enclosed spaces, such as within electrical equipment, electrical closets, or specially designed enclosures that surround a hazard. In such a case, the CO_2 system includes fixed supply, piping, and nozzles.
- Local application where the hazard can be isolated and CO_2 is applied directly on the burning material. Such a system includes a fixed supply, piping, and nozzles. System design is based on the area to be protected, nozzle design, optimum flow rates, and discharge time[1]:

Total quantity = nozzle discharge rate × number of nozzles × discharge time

- Standpipe and hand-held hoses to be directed on burning surfaces. Such a system includes supply through hoses located on reels or racks, preferably laid out so that two hoses can reach the same spot simultaneously; estimate (2 min) (500 lb/min) = 1000 lb CO_2. *Note*: The 200-ft limitation on the supply line may be extended with a bleeder, which simultaneously opens and closes a valve provided with a timer.
- Mobile systems, usually in which twin cylinders are manifolded together and installed on a dolly. Such a sys-

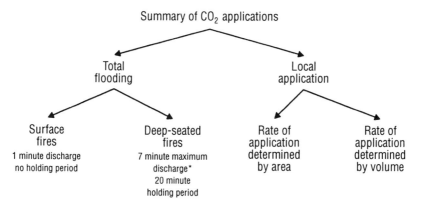

Figure 13-2 Summary of CO₂ Applications

tem is wheeled to an area where a fire is burning. Usual application is in parking garages.
- Portable fire extinguishers filled with CO_2.

Examples of CO_2 concentration for deep-seated fires include: for cable insulation, 50%; for dust-filled areas, 75%. Figure 13-2 summarizes CO_2 applications.

Advantages and Disadvantages
The advantages of CO_2 as an extinguisher are as follows:
- Provides some cooling (minor).
- Smothers fires.
- Leaves no residue after discharge.
- When used as an extinguisher, it is a gas and has the capability to penetrate and spread.

The disadvantages of CO_2 as an extinguisher are as follows:
- Hazardous to personnel in the area protected.
- Needs enclosure for best results.
- Finite supply (vs. water).
- Fire may reflash. To suppress and/or prevent reflash, provide double-shot reserve.

Alarms and Evacuation
Oxygen deficiency and decreased visibility are both concerns when CO_2 is used. For these reasons, it is important to establish an alarm system and evacuation procedure for CO_2 extinguishing systems.

There are three alarm steps in CO_2 operation: initial alarm, evacuation alarm, and discharge alarm. Each alarm has a distinctive tone: For an effective evacuation, alarm drills are required so that the occupants become familiar with the distinctive signals as well as evacuation procedures.

When CO_2 is released, auxiliary switches operated by either cylinder pressure or an electronic panel may simultaneously cut off fuel (close a gas-supply valve), close dampers, or shut off fans to cut the supply of fresh oxygen, as well as set off alarms, close fire doors, and/or shut down operating equipment.

An area protected by CO_2 must have warning signs, such as one of the following:

- Warning: Carbon-dioxide gas is discharged when alarm operates. Vacate immediately.
- Warning: Carbon-dioxide gas is discharged when alarm operates. Do not enter until ventilated.
- Warning: Carbon dioxide discharged into a nearby space may collect here. When alarm operates, vacate immediately.
- Warning: Actuation of this device will cause carbon dioxide to discharge. Before activating, be sure personnel are clear of the area.

In addition to signs, OSHA regulations require CO$_2$ discharge delays, breathing apparatus available to personnel entering the room (after the fire is out), and accessible, well-marked exits.

Specifications

For CO$_2$ system procurement, the engineer should write a specification with the idea that specialized, engineered equipment will be purchased from a vendor. Specifications must include:

- Description of the risk (hazard).
- Type of system desired (low or high pressure).
- Type of activation desired (manual and/or automatic).
- List of the opening closures to be released or activated (door fans, etc.).

The engineer must show the desired route of piping but not include sizes. Vendor drawings, together with calculations, shall be submitted for approval to the authority having jurisdiction and the owner's fire-insurance underwriter.

For final approval after installation, a puff test is usually used; however, the puff CO$_2$ discharge might not be permitted for environmental reasons. In this case, a harmless (inert) gas is used to test the system.

Cylinders and Scales

In a CO$_2$ system, high-pressure cylinders are sometimes located on a scale, which is normally inoperable unless lifted into position. Cylinders may last up to 12 years, then they must be recharged.

Table 13-1 Minimum Carbon-Dioxide Concentrations
for Extinguishment

Material	Theoretical Minimum CO_2 Concentration, %	Minimum Design CO_2 Concentration, %
Acetylene	55	66
Acetone	27	34
Aviation gas grades 115/145	30	36
Benzol, Benzene	31	37
Butadiene	34	41
Butane	28	34
Butane – I	31	37
Carbon disulfide	60	72
Carbon monoxide	53	64
Coal or natural gas	31	37
Cyclopropane	31	37
Diethyl ether	33	40
Dimethyl other	33	40
Dow therm	38	46
Ethane	33	40
Ethyl alcohol	36	43
Ethyl ether	38	46
Ethylene	41	49
Ethylene dichlorlde	21	34
Ethylene oxide	44	53
Gasoline	28	34
Hexane	29	35
Higher paraffin hydrocarbons $C_n H_{2m} + 2m - 5$	28	34
Hydrogen	62	75
Hydrogen sulfide	30	36
Isobutane	30	36
Isobutylene	26	34
Isobutylene formate	26	34
JP-4	30	36
Kerosene	28	34
Methane	25	34
Methyl acetate	29	35
Methyl alcohol	33	40
Methyl butene – I	30	36
Methyl ethyl ketone	33	40
Methyl formate	32	39
Pentane	29	35
Propane	30	36
Propylene	30	36
Quench, lube oils	28	34

Two banks of CO_2 are kept in storage for a double shot. One of the two banks of cylinders is a reserve. Cylinder weight must be checked every 6 months. If during this interval a cylinder loses 10% of its weight, it must be replaced with a new one.

Whatever the arrangement, routine maintenance should include storage-area cleanliness. Another part of routine maintenance is to ensure that all equipment is ready for proper operation when needed.

Pipe-Sizing Calculations

When CO_2 gas is discharged, the pressure drops, a vapor is formed, and CO_2 volume increases, as does friction in pipes and fittings. Computer programs are available that take all these factors into consideration and can be used when performing pipe-sizing calculations. Pipe sizing shall be done by the CO_2 manufacturer. The designer shall calculate the amount needed and select the system type (high or low pressure).

Example 13-1   Perform calculations for a total flooding system. The area in which this system will be installed contains flammable materials. Other specifications are as follows:

> Space volume: 2000 ft³
> Type of combustible: Gasoline
> Ventilation openings: 20 ft²

Solution 13-1   From Table 13-1, it is possible to determine that the design concentration of CO_2 for gasoline protection for this installation is 34%.

From Table 13-2 it is possible to determine the volume factor. For this particular installation, the room has a volume of 2000 ft³. Table 13-2 shows that between 1601 ft³ and 4500 ft³ the requirement is 18 ft³/lb CO_2. Therefore:

$$\frac{2000 \text{ ft}^3}{18 \text{ ft}^3} = 111 \text{ lb } CO_2 \text{ required}$$

Table 13-2 Flooding Factors

Volume of Space, ft³ incl.	Volume Factor		Calculated Quantity, lb, No less than
	ft³/lb CO_2	lb CO_2/ft³	
Up to 140	14	0.072	—
141 to 500	15	0.067	10
501 to 1600	16	0.063	35
1601 to 4500	18	0.056	100
4501 to 50,000	20	0.050	250
Over 50,000	22	0.046	2500

(Courtesy, NFPA no. 12)

It is necessary to account for leaks that may occur through openings. For the purposes of this example, use a quantity of 1 lb CO_2/ft² to determine the required additional amount of CO_2 needed to compensate for leaks through openings. Therefore, for a 20 ft² opening:

$$(20 \text{ ft}^2)(1 \text{ lb/ft}^2) = 20 \text{ lb}$$

The amount depends on whether the opening remains open, has a large amount of leakage, etc. For openings that are not to be closed, a calculated additional amount of CO_2 must be provided.

For this example, the total amount of CO_2 required is 131 lb (111 lb + 20 lb). Two shots are recommended, so use 131 lb × 2 = 262 lb (round to 300 lb) of CO_2, or 4 cylinders at 75 lb each. This will include 2 cylinders for the first shot and 2 for the reserve shot.

Pressure-Relief Venting Formula

Now that the total amount of CO_2 has been determined, it is necessary to calculate the size requirement for the over-pressure vent openings. For very tight spaces, there is a requirement to calculate over-pressure openings based on a pressure-relief venting formula, which is as follows:

$$X = \frac{Q}{1.3\sqrt{p}}$$

where X = Free area, in.²
Q = Calculated carbon dioxide flow rate, lb/min
p = Allowable strength of enclosure lb/ft²

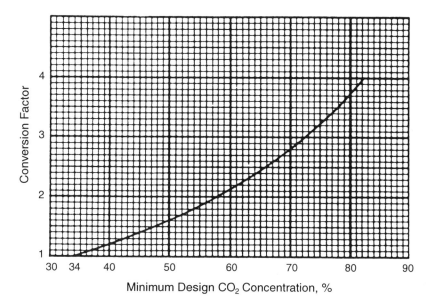

Figure 13-3 CO$_2$ Concentration

Again, this should be calculated with the manufacturer's representative's help.

Since the design requirement for this example is not more than 34% concentration, no correction factor is required for the basic quantity. If the concentration is more than 34%, the quantity of CO$_2$ required is increased by a factor of 1 to 4 (see Figure 13-3).

The pressure-relief venting factor applies to openings and is also called the "correction factor."

The amount of CO$_2$ discharged must be increased when the normal temperature of the protected space is above 200°F.

Example 13-2 Perform the calculation for a CO$_2$ installation for an electrical-equipment system.

There are two adjacent electrical-switch-gear rooms:

	Room Dimensions, ft		Area, ft^2	Height, ft		Volume, ft^3
Room no.1	(60) (60)	=	(3600)	(14)	=	50,400 ft^3
Room no.2	(60) (70)	=	(4200)	(14)	=	58,800 ft^3
Openings: 50 ft^2						

Solution 13-2 To give the preliminary estimate of CO_2 required, use the largest risk of 58,800 ft^3 and divide by the appropriate flooding factor, which can be found in Table 13-3. In this case, since the space is more than 2000 ft^3, the factor is 12 ft^3/lb CO_2. Therefore:

$$\frac{58,800 \text{ ft}^2}{12 \text{ ft}^2/\text{lb } CO_2} = 4900 \text{ lb } CO_2 \text{ required}$$

Table 13-3 Flooding Factors for Specific Hazards

Design Concentration, %	ft^3/lb CO_2	m^3/kg CO_2	lb CO_2/ft^3	kg CO_2/m^3	Specific Hazard
50	10	0.62	0.100	1.60	Dry electrical hazards in general (spaces 0 to 2000 ft^3)
50	12	0.75	0.083 (200-lb minimum)	1.33 (91-kg minimum)	Dry electrical hazards in spaces greater than 2000 ft^3
65	8	0.50	0.125	2.00	Record (bulk paper) storage, ducts, covered trenches
75	6	0.38	0.166	2.66	Fur-storage vaults, dust collectors

(Courtesy, NFPA No. 12)

Use a factor of 2 lb of CO_2/ft^2 for openings:

$$(2 \text{ lb } CO_2/\text{ft}^2) (50 \text{ ft}^2) = 100 \text{ lb of additional } CO_2 \text{ required}$$

The final amount of CO_2 required is 5000 lb (4900 lb + 100 lb). A single shot would require 5000 lb, and a double shot would require 10,000 lb. For a double-shot system (remember 2000 lb = 1 ton), use a 5-

ton, low-pressure, refrigerated tank. Using the number of cylinders required for a high-pressure system would not be a practical solution.

For gas discharge, the distance between the storage point and the discharge point is 300 ft practical maximum distance (for a low-pressure system) and 400 ft absolute maximum distance. At distances beyond these points, separate systems must be installed, with each system closer to the hazard protected.

For rotating electrical equipment, the air volume of the inside equipment to be protected must be obtained from the equipment manufacturer.

Halon Replacement

"Halon" is a generic name for a group of chemical (liquid or gas) fire extinguishers that are proven fire-suppression agents. Halons are controversial because in their current form they are damaging to the environment. Halons are part of the chlorofluorocarbon (CFC) family, and these compounds have a long atmospheric life. Ultraviolet light emitted by the sun breaks down halons, releasing chlorine and bromine into the atmosphere. These chemicals deplete the earth's protective ozone layer. At the present time, halon is replaced with a mixture of other fire-suppression gases (clear agents) with the characteristics and type of application similar to those of halon. Check with the local manufacturer's representative. NFPA 2001 describes the characteristics and qualities of halon replacement agents. (See Chapter 20 for a specification.)

Notes

[1]High-pressure cylinders use a discharge time of +30 seconds. For storage capacity, consult the vendor.

Chapter 14: Foam and Other Extinguishing Agents

Foam

Foam[1] is a smothering and cooling fire-suppression agent. It is used for extinguishing flammable and liquid, combustible fires and Class A combustibles. It is also used to prevent the release of flammable vapors and to cool sources of reignition.

Foams extinguish fires by:
- Smothering the fire by preventing the air from mixing with the flammable vapors.
- Suppressing flammable vapors and preventing their release.
- Separating flames from the fuel surface.
- Cooling fuel and adjacent (usually metal) surfaces.

Foam consists of air-filled bubbles formed from aqueous solutions, and it has a lower density than the density of any flammable liquid. Foam concentrate is produced:
- As a protein compound plus some stabilizing additives. It is diluted with water to form a 3 to 6% solution (3% foam and 97% water).
- As a fluoroprotein compound, which is similar to the one just mentioned but with a synthetic, fluorinated additive. In addition to the air-excluding blanket, foam deposits a vaporization-preventing film on the surface of a liquid fuel.

These two types of foam are used to protect areas in which hydrocarbon combustibles are involved.

Fluoroprotein bubbles are more resistant, and synthetic foam concentrates are based on foaming agents other than protein. In this group are the following types of foam:

- Aqueous-film-forming foam (AFFF), which is compatible with dry chemicals. The bubbles are lighter and break up more easily than other types. This type is used mostly for aircraft fires.
- Medium and high-expansion foams, which are used with special equipment to produce foam-to-solution volume ratios of 20:1 to 1000:1.
- Alcohol-resistant foam, or polar-solvent AFFF concentrate, which is used to fight fires on water-soluble materials, fires involving hydrocarbons, and fires involving other fuels that are destructive to AFFFs. The bubbles do not break up as quickly as those of the AFFF type.

An important element in foam-suppression systems is the correct proportion of foam solution and air. These proportions may be produced with various devices, including a fixed foam maker or a pressure foam maker. The latter device utilizes an arrangement that applies the venturi principle for aspirating air into a stream of foam solution delivered under pressure. For medium and high-expansion foam, foam generators can be either air aspirators or air blowers to introduce air into the foam solution.

Foam Application

The foam chamber is usually located at the top ring of a tank. Foam application rates range from at least 0.16 gpm/ft² up to 0.36 gpm/ft² of liquid surface area of the tank. The discharge time varies from 30 seconds to several minutes.

Foam may be applied by fixed monitors, portable monitors (on movable supports), hand-held lines, or a sprinkler system. Sprinkler systems (normally dry) that apply foam use regular, standard sprinkler heads.

The foam-suppression system is not recommended for gas fires or vertical burning fires. The foam forms a ½-in. thick blanket, which acts as an insulation between the combustible liquid and the air. Because of the high water content in foam, it cools the combustion area.

Foam is successfully used in the protection of loading docks for combustible liquids. Foam concentrate is stored in tanks with capacities between 30 and 500 gal, depending on the surface area to be protected. A foam storage tank's content must be inspected yearly, and when foam loses its quality, it must be replaced. Foam shelf life is up to 10 years.

Dry-Chemical Systems

Another fire-suppression system is the dry-chemical system.[2]

In general, dry chemicals are powder mixtures, which are used as a fire extinguisher. The dry chemicals can be applied by portable extinguishers, hand-held hose lines, or fixed (total-flooding) systems. Examples of dry chemicals include the following:

- Borax and sodium bicarbonate.
- Purple K (potassium bicarbonate).
- Super K (potassium chloride).

As a rule, dry chemicals should be used only if they have been tested and proven effective in fire suppression. *Chemicals containing oxygen (oxidizers) must not be used.* Monoammonium phosphate is a multipurpose dry chemical, which is used on wood and paper fires as well as on other hazards.

Dry chemicals are stored in pressurized containers with capacities between 30 and 300 lb. The pressurizing agent is usually nitrogen gas (inert) or carbon dioxide. Dry chemicals may be used in total-flooding or local applications. In either case, they are discharged through specially designed nozzles or hoses.

While it is somewhat unclear exactly how dry chemicals work, it is assumed that, when they are placed on a fire, the chemicals decompose and form a sticky substance that covers the fire. This isolates the fire from available, surrounding oxygen present in the atmosphere. The chain reaction is thus broken by preventing the fuel-oxygen combination reaction to continue.

Wet-Chemical Systems

Wet chemicals[3] are the same as dry chemicals, except they are mixed with water. Wet chemicals are delivered as a spray on the protected area in case of a fire. Wet-chemical fire-protection systems are designed for commercial kitchens, hoods, ducts, and associated cooking appliances. Prompt cleanup after application is recommended to minimize staining or corrosion.

Notes

[1]For more information on foam systems, see NFPA Standards nos. 11, 11A, 11C, and 16.

[2]For more information on dry-chemical systems, see NFPA Standard no. 17.

[3]For more information on wet-chemical systems, see NFPA Standard no. 17A.

Chapter 15:
Portable Fire
Extinguishers

Portable fire extinguishers[1] are classified and distinctly marked based on the type of fire on which they are to be used. For example:

- Extinguishers marked with the symbol "A" are used on combustibles such as wood, paper, and textiles. They are also used on smoldering electrical cables. These extinguishers are filled with water or multipurpose chemical agents under pressure.
- Extinguishers marked with the symbol "B" are used on combustible flammable liquids. These extinguishers may be filled with CO_2, Inergen,[2] dry chemicals, or multipurpose chemicals.
- Extinguishers marked with the symbol "C" are used on electrical-equipment fires. These extinguishers may be filled with CO_2, dry chemicals, or Inergen.
- Extinguishers with the symbol "D" are used on combustible metals, such as sodium, titanium, zirconium, and magnesium. These extinguishers are filled with special extinguishing agents.

No one type of extinguisher is effective for all types of combustible. However, extinguishers that are effective for class A, B, and C combustibles are common and used extensively.

Portable fire extinguishers are also marked with a numeral, which indicates the approximate extinguishing potential of a various type and size of extinguisher. For example, a 4A rating can put out more fires than a 2A rating. The numeral indicates the area (ft^2) of the flammable liquid that can be covered by an extinguisher.

For B extinguishers, smaller, portable sizes may be 10 to 20 lb, while larger, wheeled models may be 125 to 350 lb. For residential use, portable extinguishers smaller than 10 lb are available. See Table 15-1 for the different extinguisher ratings.

Table 15-1 Fire-Extinguisher Ratings

Combustible Type Rating	Type	Application
A	Small hose equivalent to two extinguishers	Available for quick use Indoors. Fills the same need as one or more portable extinguishers. Has the advantage of an unlimited water supply.
A	Water filled and antifreeze	Both are similar except that antifreeze has a freezing point below −40°F
B, C	Clean agent	Contains a mixture of gases super pressurized with nitrogen, which is discharged by operating a hand lever.
B, C	Carbon dioxide	Contains liquid carbon dioxide that is released by operating a valve lever or trigger. The liquid changes into gas as it Is directed on the fire through a funnel-shaped horn.
B, C	Dry chemical	Regular: Discharges a special sodium- or potassium-base powder suitable for flammable liquids and electrical equipment. Multipurpose: Discharges a powder that is suitable for ordinary combustibles, flammable liquids, and electrical equipment.
D	Dry compound	Special powders that are available for use on metal fires. They are usually applied with a shovel. A specially designed extinguisher is also available that is operated like a dry-chemical extinguisher.
B	Foam	Discharges a premixed solution of foam concentrate and water through an air-aspirator nozzle to produce foam and/or film-forming liquid.

Note: Aqueous-film-forming foam (AFFF) extinguishers are approved for use on ordinary combustibles.

A portable fire extinguisher must be conspicuously located, with its top 3 to 5 ft above the floor. Bright markings must draw attention to its location. Portable fire extinguishers are most effective when a fire just begins and there are personnel present in that area. For these reasons, portable fire extinguishers should be used only on small fires.

Regular maintenance is very important for portable fire extinguishers. The pressure gauge must be in good working condition, they must be kept clean, and they must be located at the required height, where they are easily spotted and accessible.

Portable fire extinguishers are sometimes marked with distinctive geometrical signs and colors for easy identification (see Table 15-2). (See Appendix D for a complete table of portable fire-extinguisher characteristics.)

Table 15-2 Extinguisher Characteristics

Type	Hazard	Geometrical Sign	Color*
A	Ordinary combustibles	Triangle	Green
B	Flammable liquids	Square	Red
C	Electrical fires	Circle	Blue
D	Special combustibles (metals)	Star	Yellow

* These are colors recommended by NFPA Standard no. 10 when colored, geometrical signs are used.

To sum up, portable fire extinguishers must be:
- Properly located and in good working condition.
- Conspicuously located.
- The proper type for the respective combustible material.
- Used when the fire is still small so that the extinguisher will be effective.

In addition, extinguishers must be clearly marked for easy identification, labeled, tested regularly, and inspected. Portable extinguishers must also bear the UL or ULC (Underwriters' Laboratory of Canada) label.

Portable fire extinguishers constitute the first line of defense against fires. They should be installed in addition to any other fire-extinguishing system provided. They should be located in strategic locations, including at every exit from a floor or building. In exit or access corridors, extinguishers must be located at certain distances, as indicated in NFPA Standard no. 10. They might have to be located along access or exit corridors or along the perimeter of a large, enclosed space.

A plan showing the proposed location(s) of fire extinguishers must be developed before installation. This plan must be submitted to the authorities having jurisdiction for their comment and/or concurrence.

History

The portable fire extinguisher was developed in the late 1800s. The first extinguishers consisted of a glass bottle; in 1920, the cartridge type was introduced. In 1959, extinguishers with stored water under pressure were developed, and they replaced the cartridge-operated models. In 1969, inverting extinguishers became obsolete because in some instances they were dangerous to operate. Around the same time, the vaporizing-liquid type came on the market, but it was determined that they might become poisonous, and they, too, became obsolete.

In 1973 halon 1211 was introduced to fill portable extinguishers. Today, liquefied gases such as halon 1301 are used in portable fire extinguishers. These are provided with a lever that must be squeezed to induce operation. The lever is kept in an open position by a removable pin. The use of halon in portable fire extinguishers is limited, however, because it is expensive. Also, the halon-filled extinguishers are now being replaced by extinguishers with other agents that do not contribute to ozone depletion.

Notes

[1]For more infonnation concerning portable fire extinguishers, see NFPA Standard no. 10.

[2]Inergen is a clean agent that replaces halon.

Chapter 16:
A Few
Final Words on
Fire Protection

Energy Conservation

Energy conservation is responsible for an increased amount of insulation in the building shell and a more careful seal of air leaks (in and out). These elements are good for conserving energy, but they might create problems from a fire-protection point of view. The more insulation that is installed, the more toxic fumes and smoke are produced in the case of a fire. The expected temperature of a fire is higher today because of better building insulation.

Fire Development

A fire has three periods in its development: the growth period, the fully developed period, and the decay period. Each period is characterized by the temperature of the fire and the possibility for evacuation. The fully developed fire is the most dangerous period. Each fire produces some very specific characteristics, or traces. These are known as "fire signatures."

Assessing the Life Safety in Buildings

A concern for life safety implies avoiding exposure to harmful levels of combustion products. Specific safety measures must be employed to reduce this risk. In spaces where dust is generated, dust ignition-proof equipment or an air-purged enclosure pressurized with inert gas must be provided. In such a case, the area must not be occupied, because humans could not survive. For access to such places, special protective lamps and breathing apparatus must be used.

Hazards Generated by Building Services

There are usually two major groups of building services: electrical systems and mechanical systems. These systems provide power and environmental-control conveniences and include the following:
- Lights and power.
- Air conditioning and ventilation.
- Heating.
- Refuse disposal.
- Plumbing.
- Communication.
- Transportation (e.g., elevators).
- Conveyance (e.g., escalators and conveyors).

If not properly installed, operated, and maintained, these systems may become a source of fire. While each system is independent, it is also part of the big family of building services. These systems must be adequately protected against fires by means of fixed, manual or automatic fire-protection systems as applicable. In areas where there are volatile combustibles, it is necessary to employ precautionary methods, such as explosion-proof motors, spark-proof equipment, and rubber-covered floors.

Firefighting in a High-Rise Building

A high-rise building (70 ft or taller) is a special type of structure in which firefighting and life-safety operations must be conducted from within. Local codes reflect the specific requirements for this type of structure. Specific descriptions include fire alarm and communications as well as the isolation of floor(s) where a fire develops. Smoke evacuation and smoke-free areas are of special interest. Smoke-control models are developed to assess design and future performance.

Chapter 17: Sample Specification for a Water Fire-Protection System

Water Fire-Protection System

JOB ORDER NO.:
DATE:

Table of Contents

1   **GENERAL**
1.1   DESCRIPTION
1.2   WORK PROVIDED BY THIS CONTRACTOR UNDER THIS SPECIFICATION
1.3   WORK INSTALLED, BUT EQUIPMENT FURNISHED UNDER OTHER SPECIFICATIONS
1.4   RELATED WORK PROVIDED BY OTHER SPECIFICATIONS
1.5   APPLICABLE STANDARDS
1.6   DEFINITIONS
1.7   BASE BID AND ALTERNATES
1.8   GENERAL DESIGN CRITERIA
1.9   SHOP DRAWINGS
1.10   MANUFACTURER'S DATA
1.11   AS-BUILT DRAWINGS
1.12   VALVE DIAGRAM AND OPERATING INSTRUCTIONS
1.13   CHANGES

1.14   LEAK DAMAGE
1.15   FREIGHT AND HAULING
1.16   UNIT PRICES
1.17   OVERTIME WORK
1.18   CLEANUP
1.19   SAFETY
1.20   PERMITS FROM THE AUTHORITY HAVING JURIS-
        DICTION AND FEES
1.21   GUARANTEES

2   MATERIALS
2.0   GENERAL
2.1   CABINETS
2.2   QUALITY ASSURANCE
2.3   SPRINKLERS
2.4   HOSE THREADS
2.5   SUPPLY PIPING
2.6   FITTINGS
2.7   VALVES AND DEVICES
2.8   AUTOMATIC DRY STANDPIPE SYSTEM EQUIP-
        MENT
2.9   AIR COMPRESSOR
2.10   PIPE HANGERS AND SUPPORTS
2.11   BACKFLOW PREVENTERS
2.12   ANCHORS
2.13   PIPE GUIDES
2.14   IDENTIFICATION
2.15   PRESSURE GAUGES
2.16   ACCESS PANELS
2.17   STRAINER
2.18   SLEEVES
2.19   SIGNS
2.20   FIRE DEPARTMENT CONNECTIONS
2.21   WALL INDICATOR VALVES
2.22   VARIABLE PRESSURE ALARM CHECK VALVES
2.23   WATER MOTOR GONG
2.24   PRESSURE REDUCING VALVES (PRV)
2.25   FIRE PUMPS

2.26   FIRE HOSE CABINET AND ACCESSORIES FOR BUILDING INSTALLATION
2.27   FIRE HOSE AND EXTINGUISHER
2.28   FIRE DEPARTMENT CONNECTIONS (SIAMESE)
2.29   ALARM FACILITIES
2.30   SHUT-OFF VALVE SUPERVISORY SWITCHES (TAMPER SWITCHES)
2.31   SPRINKLER CABINET
2.32   VALVE TAGS AND CHARTS
2.33   WET CHEMICAL EXTINGUISHING SYSTEMS

3   CONSTRUCTION METHODS
3.1   APPLICATION
3.2   EXCAVATING AND BACKFILLING
3.3   PIPING
3.4   OVERHEAD PIPING
3.5   TESTING AND FLUSHING UNDERGROUND PIPE
3.6   TESTING OVERHEAD PIPE
3.7   DRAINS
3.8   CEILING AND WALL PLATES
3.9   SLEEVES
3.10   FLUSHING CONNECTIONS
3.11   FIRE DEPARTMENT CONNECTIONS
3.12   ALARM VALVES
3.13   WELDING
3.14   INSPECTOR'S TEST
3.15   SPRINKLER GUARDS AND WATER SHIELDS
3.16   EXTRA SPRINKLERS
3.17   SPECIALTY DEVICES

1   GENERAL

1.1   DESCRIPTION
1.1.1   Provide all plant facilities, labor, materials, tools, equipment, appliances, transportation, supervision, and related work necessary to complete the work specified in this Section and as shown on the Drawings.

1.1.2   The work shall include careful examination of the Drawings to determine quantities, location, size, types, and details of Fire Protection work and related work described in this section.

1.1.3   The work shall be performed by a licensed Sprinkler Contractor only.

1.2   WORK PROVIDED BY THIS CONTRACTOR UNDER THIS SPECIFICATION

1.2.1   Complete automatic sprinkler equipment as shown on the Drawings.

1.2.2   New underground piping from ____ inches flange in the valve pit outside the building including all necessary fittings, clamps, blocking, valves, etc., terminating in flanged outlet plumb and level in building.

1.2.3   Provide one 2½" × 2½" × 4" flush type Fire Department connection with check valve and ball drip at sprinkler valve room.

1.2.4   Provide wet-pipe systems in building complete with control valves, alarm check valves, including trim and water motor alarm gongs, etc.

1.2.5   Provide fire hose racks complete with hose valves, fire hose, and nozzles.

1.2.6   Coordination of work with all other trades.

1.2.7   Shop Drawings including wiring diagrams.

1.2.8   Operating instructions and valve diagrams.

1.2.9   Sleeves

1.2.10 Inserts

1.2.11 Wall pipes

1.3   WORK INSTALLED, BUT EQUIPMENT FURNISHED UNDER OTHER SPECIFICATIONS

1.3.1   Installation of water flow and valve supervisory switches furnished by Alarm Contractor.

1.4   RELATED WORK PROVIDED BY OTHER SPECIFICATIONS

1.4.1   Painting of sprinkler piping and valves, including placing and removing bags or other protection devices on sprinklers to prevent paint from touching any portion of sprinkler.

1.4.2   120-V, single-phase, 60 Hz electric power supply to water flow switch.

1.4.3   Trapped floor drain in sprinkler valve room.

1.4.4   Concrete splash blocks.

1.5   APPLICABLE STANDARDS

1.5.1   Latest edition of Local and State codes as well as NFPA standards.

1.5.2   National Fire Protection Association Standards:
         A.   No. 13 – Sprinkler Systems
         B.   No. 14 – Standpipe and Hose Systems
         C.   No. 20 – Fire Pumps
         Specifics: List other standards as applicable.

1.6   DEFINITIONS

1.6.1   Contractor: The Fire Protection Contractor and any of his subcontractors, vendors, suppliers, or fabricators.

1.6.2   Provide: Furnish and install.

1.6.3   Furnish: Purchase and deliver to other trade or Owner for installation.

1.6.4   Install: Install materials, equipment, or assemblies furnished under this specification or by other trades or Owner.

1.6.5   Concealed: When used in connection with the installation of piping and accessories, it shall mean hidden from sight as in chases, furred spaces, pipe shafts, or suspended ceilings. "Exposed" shall mean not "concealed" as defined above.

1.6.6   Architect: Indicate name and address.

1.6.7   Engineer: Indicate name and address.

1.6.8   Owner: Indicate name and address.

1.6.9   UL: Underwriters' Laboratories, Inc.

1.6.10   Fire Insurance: Owner's insurance; for example, Factory Mutual Engineering Association, Insurance Services Office (ISO), or Industrial Risk Insurance (IRI), or other as applicable.

1.7   BASE BID AND ALTERNATES
1.7.1   The base bid shall be in accordance with Drawings and specifications.
1.7.2   Contractor shall state in his proposal any Contractor-proposed substitution of materials or methods of installation from that specified. These alternates shall be listed on the proposal as Contractor alternates.

1.8   GENERAL DESIGN CRITERIA
        A. All systems shall be hydraulically calculated. Hydraulic calculations signed by a registered Professional Engineer (PE) shall be submitted for approval and approved prior to starting any work. Hydraulic calculations shall be based on results of hydrant flow tests, which shall be performed by the contractor in the vicinity of each building. As a recommendation, tests shall be performed between 9:00 a.m. and 5:00 p.m. on a normal working day during summer. (Many times, conducting tests during these hours is impractical. If so, a local Fire Department Representative shall be present to "observe" the test during "off-peak hours" and to acknowledge the correctness of results.) The test results shall be submitted for review prior to submitting any hydraulic calculations. The test data shall contain the following information:
            1. Date of the test
            2. Who performed the test and who was present
            3. Site plan indicating locations and diameters of water mains and locations of the hydrants tested
            4. Grade elevations at the hydrants tested
            5. Static pressure in psi
            6. Residual pressure in psi at 1000 gpm test flow
        B. Standpipe systems shall be calculated to deliver 500 gpm to the highest and most remote hose outlet in the building with the residual pressure at 100 psi. Total flow in piping shall be equal to 500 gpm for the remote standpipe riser, plus 250 gpm for each addi-

tional riser in the same fire zone, the total not to exceed 1250 gpm.

C.* Sprinkler systems in office spaces, cafeterias, and similar occupancies shall be designed for Ordinary Group 1 Hazard, with the density 0.15 gpm/sq ft over the most hydraulically remote 2500 sq ft area. (For better protection, the owner might select larger base area and/or higher water density.)

D.* Sprinkler systems in shops (commercial space) shall be designed for Ordinary Group 2 Hazard, with the density 0.17 gpm/sq ft over the most hydraulically remote 2500 sq ft area.

E.* Sprinkler system in other sprinklered spaces shall be designed for Ordinary Group 3 Hazard, with the density 0.19 gpm/sq ft over the most remote 2500 sq ft area.

F. Area of sprinkler application for dry pipe sprinkler systems shall be increased by 30%, i.e., to 3250 sq ft.

G. Where the sprinkler system is supplied from the same piping that also supplies 2½" hose outlets, add 250 gpm inside hose allowance at the point of sprinkler piping connection to the combined piping.

H. Maximum coverage per sprinkler head shall not exceed 130 sq ft, as measured in accordance with NFPA No. 13 rules. Discharge from each sprinkler head shall not be less than required for its coverage at the density specified. Minimum pressure at any sprinkler head shall be 7 psi or that required for the minimum head discharge, whichever is greater. The spacing of sprinkler heads on the branch lines shall be essentially uniform.

I. In systems supplied directly from municipal water mains (i.e., without a fire pump), entire calculated flow shall be carried back to the connection to the city main, where additional 250 gpm shall be added

---

* The numbers shown are examples only. Specific areas for a particular job where specifications apply shall be listed.

as the outside hydrant allowance. Operating point shall be plotted on an N1.85 graph containing the hydrant flow test data.

J. In systems supplied by an automatic fire pump, the entire calculated flow shall be brought back to the fire pump discharge, and the pressure required at this point shall be compared with the pressure available at the same point when the pump capacity equals the calculated flow demand. In calculating the available discharge pressure, the pump suction pressure shall be assumed constant. It shall be calculated based on the assumed suction flow equal to 150% of the fire pump rated capacity. This suction flow shall also be applied to the city main characteristics.

K. Operation of a fire pump at 150% of its rated capacity shall not result in the pressure drop in the city main below 20 psi. The pump suction pressure shall not drop below 10 psi.

L. Friction losses in the piping shall be calculated using the Hazen-Williams formula with the "C" values in accordance with NFPA No. 13. The Darcy-Weisbach formula may also be used.

1.9   SHOP DRAWINGS

1.9.1   Within 30 days after the award of the Contract, and prior to fabrication, complete shop Drawings of the sprinkler system shall be submitted to the following:

A. Owner's Insurance Agency

B. Local and/or municipal authorities having jurisdiction

C. Architect/Engineer

1.9.2   Each shop Drawing shall bear the check stamp of the General Contractor or Construction Manager before it is submitted for the Architect's approval.

1.9.3   Prepare shop Drawing at a minimum scale of ⅛" = 1'-0" for plans, and ¼" = 1'-0" for details. Show all piping, sprinklers, hangers, flexible couplings, roof construction, and occupancy of each area, including ceiling and roof heights as required by NFPA No. 13. When welding is planned, shop Drawings shall indicate the sections to be shop welded and the type of welding fittings to be used.

1.9.4   Installation shall be based on actual survey, and all of the latest architectural, structural, heating and ventilation, plumbing, electrical drawings, and equipment installed.

1.10   MANUFACTURER'S DATA

1.10.1   Provide data from manufacturer on the following devices including installation, maintenance, testing procedures, dimensions, wiring diagrams, etc. Where any devices provided, furnished, or installed by the Contractor involve work by another Contractor, submit additional copies of data directly to that Contractor.

1.10.2   Control valves

1.10.3   Alarm check valves

1.10.4   Sprinkler heads

1.10.5   Fire department connections

1.10.6   Check valves

1.10.7   Waterflow devices

1.10.8   Valve supervisory devices

1.10.9   Fire hose

1.10.10 Valves

1.10.11 Racks

1.10.12 Nozzles

1.10.13 Water motor alarm gong

1.10.14 Water backflow preventer

1.11   AS-BUILT DRAWINGS

1.11.1   Maintain at the site an up-to-date marked set of as-built drawings which shall be corrected and delivered to the Owner upon completion of the work.

1.11.2   Furnish the Owner with one reproducible copy of each approved shop Drawing, revised to show as-built conditions.

1.12   VALVE DIAGRAM AND OPERATING INSTRUCTIONS

1.12.1   At the completion of the work, provide a small-scale plan
         of the building indicating the locations of all control valves,
         low point drains, and Inspector's test. The plans shall be
         neatly drawn and color coded to indicate the portion of the
         building protected by each system, framed under glass and
         permanently mounted on the wall adjacent to the header.
1.12.2   Furnish one copy of NFPA No. 13A and 25 bound sets of
         printed operating and maintenance instructions to the
         Owner, and adequately instruct the Owner's maintenance
         personnel in proper operation and test procedures of all
         fire protection components provided, furnished, or in-
         stalled.

1.13   CHANGES
1.13.1   Make no changes in installation from layout as shown on
         Drawings which may be requested by an Insurance Asso-
         ciation or Local Authority, unless change is specifically
         approved by the Engineer. Any changes made other than
         as above stated are at the Contractor's own expense and
         responsibility.

1.14   LEAK DAMAGE
1.14.1   The Contractor shall be responsible during the installation
         and testing periods of the sprinkler system for any damage
         to the work of others, to the building, its contents, etc.,
         caused by leaks in any equipment, by unplugged or unfin-
         ished connections.

1.15   FREIGHT AND HAULING
1.15.1   Deliver materials to job sites, unload, and store in location
         determined by Owner's representative.
1.15.2   Product Delivery, Storage, and Handling:
         Materials shall be delivered in the manufacturer's
         original, unopened, protective packages.
         Materials shall be stored in their original protective pack-
         aging and protected against soiling, physical damage, or
         wetting before and during installation.

Equipment and exposed finishes shall be protected against damage during transportation and erection. Unloading of all materials shall be done in a manner to prevent damage.

1.16	UNIT PRICES
1.16.1	Each Contractor bidding this work shall submit with his base bid, unit prices for the installation of additional sprinkler heads, complete with all necessary appurtenances. Unit prices shall be given for each different type head required in the system.
1.16.2	This price shall be based on 1 sprinkler, 1 branch line fitting, 13'-0" of branch line pipe threaded at both ends, 1 typical branch line hanger, and labor per sprinkler. Unit price for pendent sprinkler with concealed pipe shall include drop nipple, reducing couplings, and escutcheon.
1.16.3	Contractor will be required to furnish heads as required by Owner's Insurance or by Code, and no extra costs will be authorized for additional heads not shown on plans. However, if an architectural change is made which involves a change in sprinkler head count, the contract price will be adjusted up or down based on the unit price which has been agreed to.
1.17	OVERTIME WORK
1.17.1	State in bid, the extra amount to be charged for each hour of overtime work for each apprentice, fitter, foreman, supervisor, etc., that might be working on this installation.
1.17.2	Overtime work must be authorized in writing by Owner's representative.
1.18	CLEANUP
1.18.1	Maintain the premises free from accumulation of waste materials or rubbish caused by this work.
1.18.2	At the completion of the work, remove all surplus materials, tools, etc., and leave the premises clean.
1.19	SAFETY

1.19.1   All work shall be performed in compliance with the Occu-
          pational Safety and Health Act of 1970 and Construction
          Safety Act standards.

1.20   PERMITS FROM THE AUTHORITY HAVING JURIS-
          DICTION AND FEES
1.20.1   Pay all permits, fees, and charges required for this work.

1.21   GUARANTEES
1.21.1   Furnish to the Owner at the completion of this work, a
          written guarantee (in triplicate) stating that all equipment,
          materials, and work performed are in full accordance with
          the Drawings and specifications. The guarantee shall also
          state that this work and all subsequent Change Orders, are
          fully guaranteed for 1 year from the date of final accep-
          tance, and any equipment, materials, or workmanship
          which may prove defective within that time will be re-
          placed at no cost to the Owner.

2   **MATERIALS**

2.0   GENERAL
2.0.1   The naming of manufacturers in the specifications shall
          not be construed as eliminating the materials, products, or
          services of other manufacturers and suppliers having ap-
          proved equivalent items.
2.0.2   The substitutions of materials or products other than those
          named in the specifications are subject to proper approval
          of the Engineer, guaranteed in writing.
2.0.3   "Approved" shall refer to approval of the system and de-
          sign by the Authority Having Jurisdiction.
2.0.4   "Listed" shall refer to materials or equipment included in
          a list published by a nationally recognized testing labora-
          tory that maintains periodic inspection of the production
          of listed equipment or materials, and whose listing states

either that the equipment or materials meet nationally recognized standards or have been tested and found suitable for use in a specified manner.

2.1   CABINETS**

2.1.1   Extinguishers and cabinets shall be manufactured by Elkhart, Pyrene, Seco, or approved equal. Elkhart figure numbers are used to designate type and quality.

2.1.2   Cabinets shall be equal to Elkhart ST-1504, 18 gauge steel, baked white enameled inside, baked gray prime coat outside, DSA full glass door with chrome-plated door pull. Bottom of cabinets shall be set 30 inches above floor or as directed by Architect/Engineer. Cabinets shall be semi-recessed, full recessed, or surface mounted as indicated on Drawings. Extinguishers for cabinets shall be water type.

2.1.3   All extinguishers shall be furnished with wall brackets, except those in cabinets.

2.2   QUALITY ASSURANCE

All materials and equipment used in standpipe and sprinkler systems shall be UL listed.

2.3   SPRINKLERS

Sprinklers shall be used in accordance with their listings. Heads shall be fusible link or frangible bulb, spray type having ½ inch standard discharge orifice. Corrosion-resistant heads shall be used for locations exposed to either moisture or corrosive vapors, and where the installation is close to sea water. Heads shall be of ordinary temperature rating unless higher rating is indicated and/ or required by NFPA No. 13 for the areas with high ambient temperature. Sprinkler heads on exposed piping shall be upright wherever possible.

** To be used when no portable specification is included. Manufacturers and model included are just examples.

In areas with hung ceilings, piping shall be concealed with pendent heads below ceiling. Finish for heads and ceiling escutcheons shall be bright chrome.

Where pendent heads are required on dry pipe systems, the heads shall be of the special dry pendent type. Pendent heads on preaction sprinkler systems protecting heated spaces may be of regular pendent type.

Special types of heads (sidewall, extended operation, on/ off, etc.) may be utilized where their use is warranted by the specific local conditions.

Sprinkler heads located so as to be subject to mechanical injury shall be protected with approved wire guards.

The Contractor shall furnish spare automatic sprinklers in accordance with the requirements of NFPA No. 13 for stock of extra sprinklers. Spare sprinklers shall include all types and temperature ratings used and shall be furnished in a special steel cabinet which shall also house all required types of sprinkler wrenches.

2.3.1   Sprinklers for the proposed system shall be the listed automatic upright, chrome-plated pendent, or sidewall type and shall be distributed throughout the building, approximate number of sprinklers and type shown on Drawings. If the number of sprinklers indicated in the sprinkler count summary differs from actual count on plans, the actual count shall be provided.

2.3.2   Install high-temperature sprinklers of proper degree ratings wherever necessary to meet requirements of NFPA no. 13, and as follows:

A.   212°F sprinklers shall be installed as indicated and in compressor's area, computer rooms, telephone equipment room, fan room, and elevator machine room.

B.   286°F sprinklers shall be installed as indicated and in boiler rooms.

2.3.3   Install listed lead-coated or corrosive-resistant sprinklers in all areas exposed to corrosive conditions.

2.4   HOSE THREADS

2.4.1   Hose threads for hydrants and Fire Department siamese connections shall match those of the Local Fire Department.

2.5   SUPPLY PIPING
2.5.1   Sprinkler, Fire line, and Standpipe
2.5.2   All pipe, fittings, and valves shall be UL listed. Underground: All underground pipe and fittings shall be pressure Class 150 centrifugal cast-iron enameling or cement lined mechanical joint, "Tyton" joint, or approved equal. Pipe shall conform to ANSI Specification A-21.6, A-21.8, or approved equal. All underground pipe may also be ductile iron bituminous coated, cement lined with push on mechanical joints in accordance with AWWA C151, Class 51.

There are other piping materials which, if the ground characteristics require, may be used, such as: Asbestos Cement Pipe, PVC Pipe, as well as Steel Pipe. Soil analysis for corrosion protection and appropriate selection is recommended. Minimum pipe size recommended is 6" diameter.

Overhead: Overhead pipe shall be black steel and must comply with specifications of the American Society for Testing and Materials, ASTM A53, A120, A135, or A795 for black pipe, and hot dipped zinc coated galvanized welded and seamless steel pipe for ordinary uses. Galvanized pipe shall be used where exposed to atmosphere. Dimensions for all overhead pipe must be in accordance with the American Standard for wrought steel and wrought iron pipe, ANSI B-36.10-70 for pressure up to 300 psi. Schedule 40 pipe is considered "standard wall" pipe. Schedule 30 pipe is acceptable in sizes 8" and larger. The Authority Having Jurisdiction may now approve the installation of Schedule 10 pipe. This schedule pipe cannot be threaded or installed in corrosion areas. This pipe is called thin wall pipe. The same ASTM standard as for Schedule 40 applies.

Overhead pipe of the welded and seamless type specified in ASTM A53-72A used in welded systems shall have a minimum pipe wall thickness of 0.188" for pressures up to 300 psi in sizes 4" or larger. Pipe sizes 3½" and smaller, used in Schedule 10S as specified in ANSI Standard B-36-19-1965 (R1971), are permitted for pressures up to 300 psi. Pipe ends shall be rolled grooved in accordance with NFPA No. 13.

2.6   FITTINGS
2.6.1   Standard wall pipe fittings 2" and smaller shall be malleable iron conforming to ANSI B-16.3 Class 150, threaded or approved malleable iron grooved fittings and couplings similar to victaulic. Fittings 2½" and larger may be carbon steel butt-welded, factory fabricated, conforming to ANSI B-16.9 Class 150. Flanges shall be welding neck type conforming to ANSI B-16.5 Class 150.
           For thin wall pipe, listed roll-grooved fittings, welding, or listed plain end fittings shall be used.
2.6.2   Changes of direction, unless otherwise noted, shall be accomplished by the use of fittings suitable for use in sprinkler systems as defined in NFPA No. 13. Fittings exposed to the atmosphere shall be galvanized. Bushings shall not be used unless written approval is obtained from the Engineer.

2.7   VALVES AND DEVICES
2.7.1   All sprinkler control valves, devices, check valves, alarm valves, etc., shall be of the type approved and listed.
2.7.2   Gate valves 2" and smaller shall be outside screw and yoke, bronze, rising stem, wedge disk type, threaded, conforming to MSS SP-80. Gate valve 2½" and larger shall be iron body, bronze trim, outside screw and yoke, flanged, UL/FM listed conforming to MSS SP-70. All valves shall be UL listed for at least 175 psi working water pressure (wwp).
2.7.3   Globe and angle valves may be used as auxiliary valves (drain valves, test valves, trim valves, and valves on compressed air piping) for diameters not over 2".

They shall be bronze, rising stem, with bronze disk, threaded, conforming to MSS SP-80 Class 150.

2.7.4   Butterfly valves 3" and larger shall be lug style, ductile iron body, ductile iron nickel plated disk, stainless steel stem, gear operated, with a position indicator, UL listed for 175 psi wwp, conforming to MSS SP-67.

Butterfly valves 3" and larger shall not be used:

A.   On the suction side of fire pumps.

B.   Whenever the valve's position is to be supervised by a supervisory switch (tamper switch).

For diameters 2½" and smaller, if the valve requires a supervisory tamper switch, only valves listed with the supervisory attachment may be used. These valves shall have a bronze body with threaded ends, stainless steel disk and stem, visual position indicator and a built-in tamper-proof supervisory switch rated at 10 amps, 115 vac. Valve listed pressure rating shall be 175 psi wwp.

2.7.5   Check valves shall be swing type except as noted. Valves 2" and smaller shall be bronze, regrinding type with renewable disk, screwed caps, threaded, Class 150 conforming to MSS SP-80. Check valves 2½" and greater shall be iron body, bronze trim, bolted cover, flanged, conforming to MSS SP-71, UL listed for 175 psi wwp.

2.7.6   Alarm check valves shall be provided with standard variable pressure trim, including pressure gauges, retarding chamber, alarm pressure switch, main drain valve, retarding drain check valve, test apparatus, and all necessary pipe, fittings, and accessories required for a complete trimming installation, in accordance with NFPA No. 13.

2.7.7   Dry pipe valves shall be provided complete with the required trim including all required drains, gauges, testing apparatus, a water flow alarm pressure switch, high and low air pressure supervisory switches, all trim piping, and accessories. Valves supplying dry pipe systems having capacity over 500 gal, shall be provided with a quick-opening device and an anti-flooding device. Not more than 750 gal system capacity shall be controlled by one dry pipe valve. Gridded dry pipe sprinkler systems are not permitted. Water pressure on the inlet side of any dry pipe valve shall not exceed 150 psi.

2.7.8   Pre-Action (Deluge) valves shall be provided complete with the required trim, including all required drains, gauges, testing apparatus, a water flow alarm pressure switch, low supervisory air pressure alarm switch, solenoid release, control panel, supervisory air panel, and local manual emergency station.

A.   Pre-action systems shall be electrically operated and activated by cross-zoned fire detectors installed inside the protected area by others. Activation of the single detector shall result in a first detection alarm, but the pre-action valve shall not open. Operation of a second detector, installed on another electrical circuit (cross-zoned system), shall open the valve.

B.   Sprinkler piping integrity during non-fire condition shall be supervised by low pressure (max. 30 oz/sq in.) air, automatically maintained by a special compressor.

Pre-action system control panel shall be UL listed for cross-zoned operation.

All elements of the system shall be listed and be electrically compatible.

Control panel shall be provided with a standby battery power supply, rectifier, and charger. The battery shall be adequate to provide at least 24 hours of emergency power supply and be capable of opening the pre-action valve and operating all alarm signals for 5 minutes after the 24-hour standby period.

2.8   AUTOMATIC DRY STANDPIPE SYSTEM EQUIPMENT

A.   Provide deluge valves actuated by remotely located (at each hose outlet) closed circuit electric break-glass stations. Provide signs at each station giving necessary instructions for operation.

B.   Deluge valve shall be provided with all required trim including releasing system, test apparatus, alarm pressure switch, control panel with a UPS system capable of supplying entire systems with power for 24 hours, gauges, valves, and auxiliary piping.

Water flow switches shall be closed circuit paddle/vane type water flow indicators with adjustable pneumatic retarding device to prevent false alarms from line surges and required contacts and relays to activate local alarm, indicate signals on fire alarm panel in accordance with local alarm system provisions of NFPA No. 72. Pressure switches shall be 10 A, 125 vac, with tamper-proof cover, all metal enclosure, with two sets of contacts. Switches detecting operation of sprinkler system or dry standpipe systems shall operate on a pressure increase between 4 and 8 psi. High/low air pressure switches for dry pipe sprinkler systems shall be factory set to operate on increase at 50 psi and on decrease to 35 psi. *Note to specifier. Dry standpipes are one alternate. Wet standpipes may be specified as required.*

2.9   AIR COMPRESSOR

Air compressor shall be of the reciprocating air-cooled type, single stage, furnished with a filter muffler and an air check valve. Motors and starters shall be furnished.

Compressor shall be bracket mounted on the pipe or wall or floor. Each connection from the compressor to the dry pipe system shall be provided with an automatic air maintenance device containing shut-off valves, a check valve, valved bypass, strainer, pressure switch, and automatic unloader.

Provide each compressor with a pressure relief valve set at 3 psi above the compressor shut-off pressure.

Provide each compressor with an automatic controller having a Hand-on Automatic (HOA) selector switch including contacts for alarm and monitoring.

Air compressor shall have capacity sufficient to charge the system served within maximum 30 minutes from the atmospheric pressure to the normal working air pressure. If the compressor serves more than one dry pipe system, its capacity shall satisfy requirements of the largest system, and check valves and shut-off valves shall be provided on each system's air supply connection.

Air compressor shall maintain required air pressure in the dry pipe system(s) automatically.

Provide an easily replaceable cartridge-type air dehydrator for each air compressor.

2.10   PIPE HANGERS AND SUPPORTS

2.10.1   All hanger components shall be of the listed and approved type. Furnish and install hangers and supports required for piping and equipment installed. Include necessary hanger rods, as required for the work.

Piping shall be neatly and securely supported at such intervals as will prevent sagging and so that excessive loads will not be placed on any one hanger. No piping shall be supported from other piping. Install hangers at valves and equipment connections.

Maximum spacing of hangers for fire protection piping shall not exceed 10 ft and no end pipe section 2 ft or more in length shall be left without support. Distance between hangers may be increased to 15 ft for steel pipe not using vertically installed expansion shields as the structural attachments. Hanger spacing shall be less than indicated above, whenever required by NFPA No. 13 rules or by the instructions of the manufacturer of fittings and couplings.

2.10.2   Concrete Construction (5" or thicker)

     A.   For pipes 4" and smaller, use beam clamps, inserts, power driven studs, expansion bases, or Phillips type shell. Power driven studs shall be tested in accordance with NFPA No. 13, Article 3-15.

     B.   For pipes 5" and larger, use beam clamps and inserts, or in lieu of the inserts, expansion cases spaced no more than 10'-0" apart, in accordance with NFPA No. 13.

2.10.3   Concrete Construction (3" to 5" thick)

     A.   Use beam clamps or hang from top chord of joists. Do not hang from bottom chord of joists.

2.10.4   Steel Deck and Joist Construction

A. Use beam clamps or hang from top chord of joists. Do not hang from bottom chord of joist or roof deck.

2.11   BACKFLOW PREVENTERS
Backflow preventers shall be of the reduced pressure principle type conforming to the applicable requirements of AWWA C506. Furnish a certificate of approval for each design, size, and make of backflow preventer being provided for the project. The certificate shall be from the Foundation for Cross-Connection Control and Hydraulic Research, University of Southern California. It shall attest that this design, size, and make of backflow preventer has satisfactorily passed the complete sequence of performance testing and evaluation for the respective level of approval. A certificate of provisional approval is not acceptable in lieu of the above.

2.12   ANCHORS
Securely supported pipe anchors shall be furnished and installed at locations shown on the Drawings in order to obtain free expansion, contraction, and freedom from vibration under conditions of operation. Type of anchors and method of installation shall be approved prior to installation.

2.13   PIPE GUIDES
Provide pipe guides where indicated on contract Drawings and at other locations in order to maintain alignment of piping. Guides shall be approved prior to installation.

2.14   IDENTIFICATION
Each piping valve shall have a 1½" diameter brass tag with black filled engraved numbers and letter. Tag shall be affixed to valve by means of a brass "S" hook.
Tags on different services shall be identifiable by a letter and number designation.
Piping identification shall be placed in clearly visible locations. Labels and tapes of general purpose, type, and color

may be used in lieu of painting or stenciling. Colors conforming to ANSI A13.1 shall be used. Stencil piping with approved names or code letters no less than ½" high for piping and no less than 2" high elsewhere. Paint arrow-shaped markings on the lines to indicate direction of flow. Spacing of identification marking shall not exceed 50 ft. Pipe markings shall be provided for risers and each change of direction for piping.

Contractor shall install, where directed, two copies each of charts and diagrams framed and glass-covered of approved size, giving the number, location, and function of each valve and giving identification of each pipe line.

2.15   PRESSURE GAUGES

Gauges shall be provided where indicated and shall be accessible and easy to read. Gauges shall be connected by brass pipe and fittings with shut off cocks. Standoff mounting devices shall be provided for gauges for insulated piping. Gauges shall be 3½" in diameter and in accordance with ANSI B-40.1.

2.16   ACCESS PANELS

Panels shall be provided for all concealed valves, controls, or any items requiring inspection or maintenance. Access panels shall be of sufficient size and so located that the concealed items may be serviced and maintained or completely removed and replaced. Minimum size of panel shall be 12" by 12".

2.17   STRAINER

Basket or "Y" type strainers shall be the same size as the pipe lines in which they are installed. The strainer bodies shall be heavy and durable, fabricated of cast iron, and shall have bottoms drilled and plugged. Each strainer shall be equipped with removable cover and sediment basket. The basket shall be made of monel or stainless steel, with small perforations no larger in diameter than 0.033" to provide a net free area through the basket of at least 3.30 times that of the entering pipe.

2.18   SLEEVES
Pipe passing through concrete or masonry walls or concrete floors or roofs shall be provided with pipe sleeves fitted into place at the time of construction. A water-proofing clamping flange shall be installed as indicated where membranes are involved. Sleeves shall not be installed in structural members except where indicated or approved. All rectangular and square openings shall be as detailed. Each sleeve shall extend through its respective wall, floor, or roof, and shall be cut flush with each surface, and sleeves through floors and roof shall extend above the top surface at least 6" for proper flashing or finishing. Unless otherwise indicated, sleeves shall be sized to provide a minimum clearance of 1/4" between bare pipe and sleeves or between jacket over insulation and sleeves. Sleeves in water-proofing membrane floors, bearing walls, and wet areas shall be galvanized steel pipe or cast iron pipe. Sleeves in non-bearing walls, floors, or ceilings may be galvanized steel pipe or cast iron pipe.

2.19   SIGNS
2.19.1   Provide standard metal signs in accordance with NFPA No. 13.

2.20   FIRE DEPARTMENT CONNECTIONS
2.20.1   Furnish and install one 4" Model 276 Fire Department connection for each riser in location shown on the sprinkler Drawings. Connection to be furnished in bronze finish and to be as manufactured by W.D. Allen Company or equal. The complete installation shall include a check valve, ball drip, and metal escutcheon plate marked "SPRINKLER."

2.21   WALL INDICATOR VALVES
2.21.1   Recessed wall indicator valves (OS&Y gate) shall be furnished and installed on each riser. Furnish and install sleeve

for valve handle as manufactured by Traverse City Iron Works.

2.22   VARIABLE PRESSURE ALARM CHECK VALVES
2.22.1   Variable pressure alarm check valves shall be installed in each supply. The valves shall be equipped to give a signal upon operation and shall be provided with standard trimmings, including pressure gauge, retarding chamber, alarm switch, testing bypass and all necessary pipe fittings and accessories required for a completely approved installation. Provide extra auxiliary contacts for connections to central station.

2.23   WATER MOTOR GONG
2.23.1   Furnish and install an approved Water Motor Gong located on outside building walls, where directed by Architect.

2.24   PRESSURE REDUCING VALVES (PRV)
2.24.1   Whenever maximum pressure can exceed 175 psi in the wet pipe sprinkler system or 150 psi at any dry pipe valve, provide approved pressure reducing valves to reduce static and residual pressure to the acceptable level. Valves shall be selected based on a careful analysis of the system flow/pressure demand versus supply flow/pressure characteristics within the entire range of the possible supply fluctuations. This data shall be plotted on the PRV curve and submitted for approval for each particular PRV location.

2.25   FIRE PUMPS
           A.   The fire pumps shall be electrically motor driven and shall be horizontal split case, double suction, bronze fitted, centrifugal type installed in accordance with the requirements of NFPA No. 20. Pumps shall be furnished complete with all fittings and accessories as required by NFPA No. 20. Pump and motor shall be mounted on a rigid one piece cast iron, drip rim base, grouted in accordance with Hydraulic Institute Standards.

B. The pumps shall deliver no less than 150 percent of rated capacity at a pressure no less than 65 percent of rated pressure. The shut-off pressure shall not exceed 140 percent of rated pressure.

C. The electric motor shall be a horizontal, floor mounted, open drip proof (or TEFC), ball bearing type, ac, squirrel cage induction motor with strip heater. Locked rotor current shall not exceed the values specified in NFPA No. 20.

D. The fire pumps shall be UL listed.

E. The fire pump controller shall be a factory assembled, wired, and tested unit. Controller shall include contacts for remote control and monitoring.

F. The automatic transfer switch shall be an integral part of the fire pump controller, UL listed, rated as required to satisfactorily operate the fire pump motor. The construction of the automatic transfer switch shall conform to UL 1008.

G. The fire pump shall be painted or enamel finished with the manufacturer's approved red paint for this application.

H. The fire pump manufacturer shall factory test hydrostatically and run test for each pump prior to shipment. The pump shall be hydrostatically tested at a pressure of no less than 1½ times the no flow (shutoff) head of the pump's maximum diameter impeller plus the maximum allowable suction head, but in no case less than 250 psi. These data shall be submitted and shall include a Certified Pump Curve.

I. Each electrical fire pump shall be electrically supervised to indicate the following operating conditions and alarms:

1. Pump is in operation

2. Power failure

3. Phase reversal on the line side of the motor starter

4. Trouble alarm shall be generated in the event of inadvertent circuit opening or grounding

J.   The manufacturer's qualified representative shall participate in a field acceptance test upon completion of each pump installation. The test shall be made by flowing water through calibrated nozzles, approved flow meters, or other accurate approved devices. The test shall be conducted as recommended in NFPA No. 20 by the Contractor. Failure to submit documentation of factory and field tests will be just cause for equipment rejection.

K.   A jockey (pressure maintenance) pump shall be provided for each fire pump installation. The jockey pump shall be of the vertical or horizontal type, with rated capacity not less than the normal leakage rate of the system served and with the discharge pressure sufficient to maintain the desired fire protection system pressure. The pump capacity-pressure curve shall not be steep.

L.   Jockey pump controller shall be of the type specifically manufactured to control jockey pumps in fire-protection systems and shall be equipped with a minimum running period timer and contacts for remote control, monitoring, and alarm.

2.26   FIRE HOSE CABINET AND ACCESSORIES FOR BUILDING INSTALLATION

2.26.1 Fire Hose Cabinets in Finished Building Spaces
Fire hose cabinet shall be full recess, 18 gauge steel box with baked white enamel interior, 20 gauge tubular steel door with 16 gauge frame and with a continuous steel hinge (brass pin) and full panel wire glass. Steel corner seams shall be welded and ground smooth. Door and frame shall be finished with baked on gray prime coat. Cabinet shall accommodate a single 2½" by 1½" fire hose rack assembly with hose, an aluminum 1½" by 2½" spanner wrench, bracket mounted inside the cabinet, and a portable, 10 lb, ABC-type fire extinguisher.
Rack assembly shall be 2½" by 1½" hose rack unit with the following items:

1. Valve: 2½" double female threaded, polished chrome-plated, 300 psi, UL labeled angle valve.
2. Valve shall be provided with an adjustable pressure restricting device where the maximum pressure exceeds 100 psi.
3. Hose Rack: For use with 2½" hose valve and 100 ft of specified 1½" hose, polished chrome-plated steel, with hose pins and water retention device.
4. Rack Nipple: 2½" polished chrome-plated steel.
5. Adjustable 2½" pressure restricting device (where required), chrome-plated, furnished with field setting chart.
6. Reducer: 2½" by 1½" polished chrome-plated steel, female inlet, male outlet.
7. Hose: 1½" single jacket, polyurethane lined, 100% synthetic, rated for 500 psi test. The hose shall have UL and FM labels. Hose on each rack shall consist of no more than two single lengths connected by couplings.
8. Hose couplings: 1½" polished chrome-plated, steel, pin lug type.
9. Hose Nozzle: Polished chrome-plated, with rubber bumper, three positions: shut-off, straight stream, and fog.
10. Escutcheon Plate: 2½" chrome-plated steel.
11. Identification: Identifying decal on the cabinet's glass door shall read: "FIRE EXTINGUISHER."
12. Alternative: Two separate hose connections, one 2½" and one 1½". All other items as applicable.

2.27   FIRE HOSE AND EXTINGUISHER

2.27.1   Fire Hose Cabinets in Unfinished Building Spaces
Fire hose cabinets shall be surface mounted. They shall be equal to the cabinets specified for finished building spaces except as noted. Valves, nipples, reducer, nozzles, and couplings shall be brass finish. Hose rack shall be finished in

red baked enamel. Cabinet shall be finished in baked white enamel.

2.28   FIRE DEPARTMENT CONNECTIONS (SIAMESE)

2.28.1   Fire department connections shall be in accordance with the Fire Department equipment and NFPA No. 13. Siamese size and type shall be as indicated on the plans. Their characteristics:

1. Finish shall be dark statuary bronze.
2. Inlets shall be equipped with clappers and cast brass caps with brass chains.
3. Provide each Siamese with a statuary bronze wall plate with cast-in words indicating protected area and type of system: "STANDPIPE," "DRY STANDPIPE," "COMBINATION STANDPIPE AND SPRINKLER SYSTEM," etc.
4. Inlet threads to match fire department standard.
5. Where required to prevent freezing, provide ¾" automatic ball drips, installed horizontally, and spill water to the appropriate locations.

2.29   ALARM FACILITIES

2.29.1   Provide alarm initiating devices as follows:

1. Tamper switches: for all shut-off valves controlling water supply to sprinklers.
2. Waterflow alarm switches: for each sprinkler connection to riser with more than 20 sprinkler heads supplied by the connection.
3. Fire alarm pressure switches: for each alarm check valve, dry pipe valve, and pre-action valve installation.
4. Low air pressure and high air pressure switches: for each dry pipe valve installation.
5. Low supervisory air pressure switch for each pre-action valve installation.
6. Loss of regular power supply to pre-action system.
7. Trouble alarms.

2.30   SHUT-OFF VALVE SUPERVISORY SWITCHES (TAMPER SWITCHES)

A.   Tamper switches shall be provided on all valves controlling water supply to sprinklers, both on sprinkler piping and on combined standpipe/sprinkler piping.

B.   Valves controlling water supply to standpipe only shall be secured in fully open position by a chain and a lock.

2.31   SPRINKLER CABINET

Sprinkler cabinet for storage of spare sprinkler heads shall be enameled steel finished in baked red enamel. Furnish for each building a cabinet with sprinkler heads of all types and temperature ratings used and sprinkler head wrenches. Number of spare sprinklers shall be in accordance with NFPA No. 13.

2.32   VALVE TAGS AND CHARTS

The Contractor shall provide separate charts or diagrams and lists showing the essential features of the sprinkler and standpipe systems, including all equipment, valves, and controls, which shall be located and designated by numbers, corresponding to the numbers, stamped and painted on 2" diameter aluminum tags fastened with a brass hook on all valves (not to the hand wheels). Charts, diagrams, and lists shall be of approved size and type and mounted in glazed frames, securely attached to the wall, where directed.

2.33   WET CHEMICAL EXTINGUISHING SYSTEMS

2.33.1   Pipe shall be black steel ASTM-A53 with threaded or welded connections. If welded connections are used, the pipe wall thickness shall not be less than 0.188" or Schedule 40 pipe thickness, whichever is greater. If threaded connections are used, pipe wall thickness shall be in accordance with Schedule 40.

A.   Pipe-thread compound or tape shall not be used.

2.33.2   Fittings shall be malleable iron threaded, Class 150, ANSI
          B-16.3, or factory-made, standard weight, wrought seam-
          less steel welding fittings ANSI B-16.9.
2.33.3   Exposed pipe and fittings shall be chrome-plated.
          A. Galvanized pipe and fittings shall not be used.

3   CONSTRUCTION METHODS

3.1   APPLICATION
          A. Installation
              1. Materials and equipment shall be installed in
                  accordance with the manufacturer's recommen-
                  dations, NFPA No. 13, NFPA No. 14, NFPA No.
                  20, and the last edition of any other standard
                  applicable.
              2. Install standpipe and sprinkler piping so that the
                  entire system may be drained. Minimum pitch
                  to drains in dry pipe systems shall be in accor-
                  dance with NFPA No. 13. Wet pipe system pip-
                  ing may be installed level, with each trapped
                  portion provided with an auxiliary drain. Type
                  of auxiliary drain shall be in accordance with
                  NFPA No. 13.
              3. Cutting structural members for passage of pipes
                  or for pipe hangers fastenings shall not be
                  permitted.
              4. Pipe suspension from ceiling shall be from beam
                  clamps, steel fishplates, cantilever brackets, or
                  by rods with double nuts. Additional steel fram-
                  ing shall be provided where necessary. Supports
                  shall secure piping in place, maintain required
                  pitch, prevent vibration, and provide for expan-
                  sion and contraction.
              5. Vertical piping shall be supported by extension
                  pipe clamps bolted on each side of pipe and bear-
                  ing equally on structure or welded to beam.
                  Spacing shall be one support for each pipe sec-
                  tion.

6.   Use extra heavy pipe for nipples where unthreaded portions of pipe are less than 1½" long. Closed nipples shall not be used.

7.   Provide elbow swings or expansion loops in piping for vertical and horizontal expansion and contraction and building expansion joint crossings.

8.   Pipe 4" and larger shall be supported with clevis hangers, under 4" galvanized solid band, adjustable ring-type hangers may be used. Wall brackets shall be used for wall supported pipe. Floor mounted pipe shall be on pipe saddles.

9.   Drains and test pipes

B.   Drains shall be as follows:

1.   Siamese drain shall have ¾" auto ball drips.

2.   Other drains shall have valves, plugs, or both as required.

3.   Drains shall be provided at the following:

a.   Between a Siamese and a check valve

b.   At base of risers

c.   On valved sections

d.   At alarm, dry, and deluge valves

e.   At other required locations for complete drainage of system

C.   Test Pipes:

1.   Test pipes shall be minimum 1" pipe size, valved, and piped to discharge through the proper orifice.

2.   Test pipes shall have a sight glass, unless the water discharge can be observed from the test valve location.

D.   Underground pipe:
Clamp and block all underground piping where required and in accordance with the requirements of NFPA No. 24.

3.2   EXCAVATING AND BACKFILLING
3.2.1   General

A. Perform all excavation, including necessary shoring, and all backfilling required for the completion of work under this contract that is to be installed underground, outside, or within building walls. The arrangement of shoring shall be such as to prevent any movement of the trench banks and consequent strain on the pipes.

B. Place all surplus dirt where directed by the Engineer.

3.2.2   Excavation:

A. Excavate to the required depth and grade to the bottom of the trench to secure the required slope.

B. Rock or concrete where encountered, shall be excavated to a minimum depth of 6" below the bottom of pipe.

C. Submit certificates from a testing laboratory certifying that the backfilling and compaction thereof is in accordance with the requirements, before final pavement is installed.

D. Where mud or otherwise unstable soil is encountered in the bottom of the trench, such soil shall be removed to firm bearing and the trench shall be backfilled with sand to the proper grade and tamped to provide uniform firm support.

E. Pipe shall not be laid on frozen subgrade.

F. Minimum pipe cover where no freezing occurs shall be 2½ ft. Where freezing occurs, it shall be below freezing line for the area.

3.3   PIPING

3.3.1   Pipe shall be thoroughly cleaned after cutting or flame beveling. Pipe ends shall be square, reamed, and then threaded or flame beveled when welded. Threads shall be free of burrs. During construction, open ends of piping shall be protected with temporary closure plugs.

All pipe ends shall be reamed to full size and all threads shall be clean cut. Joints in screwed piping shall be made with joint compound or Teflon tape.

Ends shall be beveled before welding. Welds shall be made without backing rings. Welds shall be clean and free of metal "icicles," loose metal, or other obstructions that can result from welding and retard flow of gases and fluids. When a leak appears in a weld, do not re-weld without first chipping away the weld of about ½" each way, and then weld over the chipped surface. Slip-on flanges shall be welded both front and back.

3.4   OVERHEAD PIPING

3.4.1   All sprinkler piping, drain and test piping, Fire Department connection piping, etc., installed through exterior walls shall be galvanized. All sprinkler piping must be substantially supported from building structure and only approved type hangers shall be used. Sprinkler lines under ducts shall not be supported from duct work but shall be supported from building structure with trapeze hangers where necessary or from steel angles supporting duct work in accordance with NFPA No. 13. Do not hang from bottom chord of joists. Hang from top chord only, at panel point.

In addition, pipes 4" and larger, when running parallel with joists, are to be supported from trapeze hangers from 2 bar joists. If pipes run parallel with a beam, they can be hung from the bottom flange of the beam. If pipes 4" and larger are supported from top chords with hangers they shall be spaced a maximum of 10'-0" on center.

3.4.2   Sprinklers below ceilings which are on exposed piping shall be listed and approved regular bronze upright type, in upright position, except listed and approved regular bronze pendent type, in pendent position, may be used on wet pipe systems where necessary, due to clear height requirements, duct interferences, etc.

3.4.3   Pendent sprinklers under plastered, gypsum board, or acoustical suspended ceilings shall be listed and approved, chrome-plated pendent type with chrome-plated ceiling plate of 1" maximum depth with supply piping concealed above ceiling. Pendent sprinklers below ceiling shall be in alignment and parallel to ceiling features, walls, etc. Adjustable ceiling plates shall not be used unless specified.

Install sprinkler piping in exposed areas as high as possible using necessary fittings and auxiliary drains to maintain maximum clear head room.

3.4.4   Complete sprinkler equipment and place in service during nonworking hours in all areas where merchandise or fixtures are stored or in place. Provide sprinkler protection before combustible contents are moved into building.

3.4.5   Install paired flanges and numbered test blanks to provide partial protection during construction. Maintain a "test blank log," as shown on Drawings, at the site during construction to ensure removal of all blanks at completion of job.

3.5   TESTING AND FLUSHING UNDERGROUND PIPE

3.5.1   Test all new underground piping for a period of 2 hours at a hydrostatic pressure of 200 psi in accordance with NFPA No. 24, and leakage <u>shall not</u> exceed 2 quarts per hour per 100 joints. If repairs are made, retest after such repairs are completed.

3.5.2   Test shall be made before trench in which pipe is laid is backfilled.

3.5.3   Flush all underground piping thoroughly in accordance with the requirements of NFPA No. 13, Article 1-11.2, and flush test must be witnessed by proper authority.

3.5.4   Underground piping shall be flushed before connection to interior sprinkler piping is made.

3.6   TESTING OVERHEAD PIPE

3.6.1   Test all overhead sprinkler piping for a period of 2 hours, at a hydrostatic pressure of 200 psi and all piping, valves, sprinklers, etc., shall be watertight. Notify Owner's and Engineer's representative 48 hours in advance regarding time and date of all tests.

3.6.2   Performance Test Reports
         Upon completion and testing of the installed system, test reports shall be submitted in booklet form showing all field tests performed.

3.7	DRAINS
3.7.1	Provide 2" main drain valves at system control valves and extend piping to 6" open hub drain.
3.7.2	Pipe all drains and auxiliary drains to locations where water drained will not damage stock, equipment, vehicles, planted areas, etc., or injure personnel.
3.7.3	Plugs used for auxiliary drains shall be brass.
3.7.4	All piping and fittings downstream of drain valve shall be galvanized.

3.8	CEILING AND WALL PLATES
3.8.1	Install chrome-finished ceiling and wall plates wherever exposed sprinkler piping passes through ceilings and walls.

3.9	SLEEVES
3.9.1	Set sleeves in place for all pipes passing through floors and masonry walls opening.
3.9.2	Space between sleeve and pipe shall be filled with non-combustible packing.
3.9.3	Sleeves through floors shall be watertight.

3.10	FLUSHING CONNECTIONS
3.10.1	Provide flushing connections in cross mains as specified in NFPA No. 13.

3.11	FIRE DEPARTMENT CONNECTIONS
3.11.1	Install Fire Department connection properly connected to piping with necessary check valve and ball drip drain connection.
3.11.2	Provide standard nameplate marked: "AUTOMATIC SPRINKLERS"

3.12	ALARM VALVES
3.12.1	Install 8" alarm check valves, complete with trimmings including retarding chambers connected to outside water motor alarm gongs.

3.13	WELDING

3.13.1   No field welding of sprinkler piping shall be permitted.

3.13.2   Join all inside piping by means of screwed, flanged, or flexible gasketed joints or other acceptable fittings.

3.13.3   Cross mains and branch lines may be shop welded using acceptable welding fittings with screwed branch outlets. Welding and brazing shall conform to ANSI B-31.1, 1967, with latest Addenda ANSI B-31.1a and ANSI B-31.1b, 1971. Welding and torch cutting shall not be permitted as a means of installing or repairing sprinkler systems.

      A.   Provide a blind flange at each end of welding header.

3.13.4   Certify welders or brazers as being qualified for welding and/or brazing in accordance with the requirements of ASME Boiler and Pressure Vessel Code, Section IX, Qualification Standard for Welding and Brazing Procedures, Welders, Brazers, and Welding and Brazing Operators, latest edition.

3.14   INSPECTOR'S TEST

3.14.1   Provide inspector's test connections, as specified in NFPA No. 13 at required points for testing each water flow alarm device. Special discharge nozzle shall have same size orifice as majority of sprinklers installed.

3.14.2   Provide 1" sight glass where inspector's test discharge cannot be readily observed while operating valve.

3.14.3   Pipe all inspector's test connections discharging to atmosphere to location where water drained will not damage stock, equipment, vehicles, planted areas, etc., or injure personnel.

3.14.4   All pipe and fittings downstream of inspector's test valve shall be galvanized.

3.14.5   Consult with Architect's representative at job for exact location of inspector's test connections.

3.15   SPRINKLER GUARDS AND WATER SHIELDS

3.15.1   Provide guards on sprinkler within 7'-0" of finished floor or wherever sprinklers may be subject to mechanical damage.

3.15.2   Provide water shields for sprinklers installed on high shelves, under grating, and as required.

3.16   EXTRA SPRINKLERS
3.16.1   Provide spare sprinkler cabinets complete with sprinklers of assorted temperature ratings of the type necessary and in use throughout the installation, including 2 special sprinkler wrenches.
3.16.2   Install sprinkler cabinet at the wall near sprinkler risers.
3.16.3   Confer with Architect's representative for exact location of cabinet.

3.17   SPECIALTY DEVICES
3.17.1   Installation of all specialty devices shall be in accordance with manufacturer's instructions. Where the installation of those devices require the use of a torque wrench or other appliance, the Contractor shall certify that the manufacturer's instructions have been complied with and one special wrench or tool shall be left for the Owner's maintenance personnel's usage.

Chapter 18: Sample Specification for Portable Fire Extinguishers

1 SUMMARY

The work of this section consists of providing fire extinguishers and fire extinguisher mounting brackets.

1.1 APPLICABLE PUBLICATIONS

The publications listed below form a part of this specification to the extent referenced. Specifications are referred to in the text by basic designation only. In case of conflict between provisions of codes, laws, ordinances, and these specifications, including the Contract Drawings, the most stringent requirements will apply.

A. National Fire Protection Association (NFPA) Standards:
NFPA No. 10 – Portable Fire Extinguishers[1]

B. Underwriters' Laboratories, Inc. (UL) Publications:
Fire Protection Equipment Directory
ANSI/UL 154 – Carbon Dioxide Fire Extinguishers[2]
ANSI/UL 299 – Dry Chemical Fire Extinguishers[3]

1.2 GENERAL REQUIREMENTS

A. All fire extinguishers shall be UL listed. Material and equipment shall be the standard product of a manufacturer regularly engaged in the manufacture of such product and shall essentially duplicate equipment that has been in satisfactory service for five years.

B. Nameplates:

Fire extinguishers shall have a securely affixed standard nameplate showing the manufacturer's name, address, type or style, model, serial number, catalog number, and UL listing mark.

1.3   SUBMITTALS
Submit manufacturer's standard drawing or catalogue cuts of the equipment and methods of installation.

1.4   PRODUCT DELIVERY, STORAGE, AND HANDLING
A.   Materials shall be delivered in the manufacturer's original unopened protective packages.
B.   Materials shall be stored in their original protective packaging and protected against soiling, physical damage, or wetting before and during installation.
C.   Extinguishers shall not be installed on brackets until the entire installation is completed and ready to be turned over to the owner.
D.   Equipment and exposed finishes shall be protected against damage and stains during transportation and erection.
E.   Unloading of all materials shall be done in a manner to prevent misalignment or damage.

2   **MATERIALS**

2.1   GENERAL
Shall be an Underwriters' Laboratories approved, multipurpose, stored pressure, dry chemical unit approved for use on Class A, B, and C fires. The extinguisher shall have trigger-type valves with hand grips and be equipped with a short hose and diffuser horn and spring clip wall mounting bracket. The unit shall be a 10-lb[4] unit having minimum Underwriters' Laboratories ratings of 2A: 20-B:C and shall be a No. _____ as manufactured by W.D. Allen Mfg. Co., Broadview, Ill.; General Fire Extinguisher Corp., Northbrook, Ill.; Elkhart Brass Mfg. Co., Elkhart, Ind.; or an acceptable equivalent product.

2.2   PORTABLE EXTINGUISHERS AND BRACKETS
    A.   Shall be hand-portable, charged, and rechargeable, with pull ring lock pin on squeeze handle, and wall mounted or stored in fire hose cabinet. Portable fire extinguishers shall conform to NFPA Standard No. 10. All extinguishers shall be new.
    B.   Dry Chemical Extinguishers:
       Multi-purpose, ABC class powder base, designed, constructed, tested, and marked in accordance with UL 299; capacity of 20 lb, minimum rating ____.
    C.   Carbon Dioxide Extinguishers:
       Designed, constructed, tested, and marked in accordance with UL 154; capacity to 10 lb, minimum rating 1OBC.
    D.   Bracket:
       Heavy duty, anchored in place, quick release type. Or as an alternative to materials use portable fire extinguishers.

3   CONSTRUCTION METHODS

3.1   APPLICATION
    A.   Extinguishers are to be conspicuously located, readily accessible, and immediately available for use.
    B.   Provide fire extinguishers spaced in accordance with the following criteria:
       1.   For Class A hazards, extinguishers shall be located throughout the protected area, and travel distance from any point to the nearest extinguisher shall not exceed 75 ft.
       2.   Carbon Dioxide—Carbon dioxide fire extinguishers shall be provided where a fire either directly involving or surrounding electrical equipment can happen.
       3.   Maximum floor area per extinguisher shall not exceed 11,250 sq ft.

    4.   For Class B hazards, travel distance from any point of the protected area to the nearest extinguisher shall not exceed 50 ft.

  C.   Portable extinguishers are to be maintained fully charged and kept in their designated places at all times when not in use.

Notes

[1] Latest edition.

[2] *Ibid.*

[3] *Ibid.*

[4] The capacity extinguisher numbers shown are for example only.

Chapter 19: Sample Specification for a High-Pressure CO$_2$ Fire-Protection System

1   **SUMMARY**
The work specified in this section consists of furnishing, installing, and testing complete carbon dioxide fire suppression systems including all required fire detectors, local audio and local alarm wiring, and self-contained breathing apparatus.

1.1   APPLICABLE PUBLICATIONS
The publications listed below form a part of this specification to the extent referenced. The publications are referred to in the text by basic designation only. In case of conflict between provisions of codes, laws, ordinances, and these specifications, including the contract Drawings, the most stringent requirements shall apply.
    A.   National Fire Protection Association (NFPA) Standards (latest edition as applicable):
        Carbon Dioxide Extinguishing Systems (12)
        National Fire Alarm Code (72)
    B.   Underwrites' Laboratories, Inc. (UL) Publication:
        Current Fire Protection Equipment Directory

1.2   GENERAL REQUIREMENTS:
    A.   Standard Products

Material and equipment shall be the standard products of a manufacturer regularly engaged in the manufacture of fire protection products for a minimum period of five years.

1. Activation of the first detector of a cross-zoned pair shall result in local audio/visual detection alarms.

2. Activation of the second detector of a cross-zoned pair shall activate local audio/visual evacuation alarms. Simultaneously, a signal shall be transmitted to the local fire alarm control panel. It shall also close hvac dampers, shut off the fans serving the protected area, and start a discharge timer (30 seconds to 2 minutes, adjustable), which will provide carbon dioxide discharge upon expiration of the time delay. Note to specifier: Other arrangements shall be described to fit the job.

B. Verification of Dimensions

The contractor shall become familiar with all details of the work, verify all dimensions in the field, and report any discrepancy before performing the work.

C. Welding

Welding shall be in accordance with the qualified procedures and in accordance with NFPA 12. Only shop welding with appropriate welding fittings shall be used.

1.3   SUBMITTALS

A. Shop Drawings:

Shop drawings shall include performance charts, instructions, catalog information, diagrams, and complete shop drawings of the carbon dioxide system including all hydraulic calculations and other information required to illustrate that the system complies with all requirements of the Contract Documents and will function as a unit. Shop drawings shall conform to the requirements established for working plans as prescribed in NFPA 12 and shall contain complete wiring and interconnection diagrams and detailed description of the sequence of operation.

B. Test Procedures:
The Contractor shall furnish detailed test procedures with test forms for the fire protection system prior to performing system tests.

C. Spare Parts Data:
After approval of the shop drawings, the Contractor shall furnish spare parts data for each different item of material and equipment specified.

D. Operating and Maintenance Instructions:
1. Contractor shall furnish six complete copies of operating and maintenance instructions outlining the step-by-step procedures required for system start up, operation, and shutdown.
2. Contractor shall furnish framed instructions in laminated plastic and shall include wiring and control diagrams showing the complete layout of the entire system.

E. Performance Test Reports:
Upon completion and testing of the installed system, test reports shall be submitted in booklet form showing all field tests performed.

F. Calculations:
Calculations of the required amount of carbon dioxide, pipe sizes, and nozzle orifice sizes shall be signed by a registered Professional Engineer, submitted for approval, and approved prior to starting any work.

1.4   DESIGN CRITERIA
A. As an example, carbon dioxide systems shall be designed for protection of under floor spaces in computer rooms.

B. Carbon dioxide shall be supplied from high pressure (850 psi) cylinders. Amount of carbon dioxide in cylinders shall be calculated based on 75% concentration of gas in accordance with all applicable requirements of NFPA 12.

C. Design concentration in the protected space shall be achieved within a maximum of 1 minute.

D. Additional quantity of carbon dioxide shall be provided to compensate for any special condition that may adversely affect the extinguishing efficiency. Any opening that cannot be closed at the time of extinguishment shall be compensated for by the addition of a quantity of carbon dioxide equal to the anticipated loss at the design concentration during a 1-minute period. This amount of carbon dioxide shall be applied through the regular distribution system.

E. Reserve cylinders and a main-to-reserve switch shall be provided to ensure uninterruptible protection.

F. Carbon dioxide systems shall be electrically activated by cross-zoned fire detectors located within the protected space. Two detectors located on separate electric circuits shall be activated to cause carbon dioxide discharge.

G. Carbon dioxide shall also be discharged upon activation of a pre-action system, if such system is installed, to protect the above floor space, regardless of the status of the under floor detectors. Opening of the pre-action valve shall start carbon dioxide discharge timer, activate evacuation alarms, close hvac dampers, and shut off the fans serving the protected area.

H. Design shall incorporate all safety requirements in accordance with NFPA 12, with special attention to fire detection alarms, pre-discharge alarms, signage, breathing apparatus, and personnel training.

I. Manual discharge controls shall be provided in addition to automatic means. These controls, when activated, shall cause the complete system to operate in its normal fashion and shall not cause the time delay to recycle. These shall be combination normal/emergency controls, complying with NFPA 12 requirements for both types.

J. Entire system shall be electrically supervised.

1.5   PRODUCT DELIVERY, STORAGE, AND HANDLING
        A.   Materials shall be delivered in the manufacturer's
              original unopened protective packages.
        B.   Materials shall be stored in their original protective
              packaging and protected against soiling, physical
              damage, or wetting before and during installation.
        C.   Equipment and exposed finishes shall be protected
              against damage during transportation and erection.
        D.   Unloading of all materials shall be done in a manner
              to prevent damage.

2   **MATERIALS**

2.1   QUALITY ASSURANCE
        All materials and equipment used in carbon dioxide fire
        suppression systems shall be UL listed.

2.2   PIPE, FITTINGS, AND ACCESSORIES
        A.   Pipe shall be galvanized black steel Schedule 80
              ASTM A-53 seamless or electric welded, Grade A or
              B, or ASTM A-106, Grade A, B, or C. Furnace butt
              weld ASTM A-53 pipe shall not be used. ASTM A-
              120 or ordinary cast iron pipe shall also not be used.
        B.   Fittings shall be screwed or flanged, galvanized, Class
              300 malleable or ductile iron for diameters up to and
              including two inches, forged steel fittings for larger
              sizes. Flanged joints upstream of any stop valves shall
              be Class 600 and downstream of any stop valve or in
              systems with no stop valves may be Class 300.
        C.   The piping system shall have a minimum bursting
              pressure of 5000 psi.

2.3   VALVES
        A.   All valves shall be suitable for intended use, particu-
              larly in regard to flow capacity and operation.
        B.   Valves shall have a minimum bursting pressure of
              6000 psi if they are constantly under pressure and
              5000 psi if they are not under constant pressure.

2.4   NOZZLES
    A.   Nozzles shall be listed for their discharge character-
        istics and shall consist of an associated horn, shield,
        or baffle.
    B.   Discharge nozzles shall be made of corrosion-resis-
        tant metal and shall be of adequate strength for use
        with the expected working pressure and constructed
        to withstand expected temperatures without deforma-
        tion.
    C.   Nozzles shall be permanently marked to show the
        equivalent "Standard" orifice size as per NFPA 12.

2.5   CYLINDER BATTERY
    A.   Mount cylinders in normal upright position.
    B.   Connect outlets with flexible metal loops to the mani-
        fold.
    C.   Cylinders shall be equipped with brass cylinder valves
        having siphon tubes extended from valves to bases
        of cylinders and safety disks.
    D.   Where one carbon dioxide installation protects more
        than one hazard with the use of directional valves,
        the amount of carbon dioxide in the cylinder battery
        shall be adequate to protect the largest of the pro-
        tected volumes.
    E.   Provide an equal reserve battery, manifolded with
        main battery, with quick changeover arrangement, so
        that automatic and manual releases can be immedi-
        ately transferred to operate the reserve battery.

2.6   DIRECTIONAL VALVES
    A.   Directional valves shall automatically direct gas to
        the proper space when several spaces are protected
        from a single battery or cylinder manifold.
    B.   Directional valves shall be manifolded together and
        connected to the cylinder manifold.
    C.   Valves shall be bronze, threaded.

2.7   MANUAL PULL STATIONS

    A.   Manual stations shall be mechanically or electrically operated break-glass boxes located above floor within and outside of the protected room.

    B.   Activation of a manual station shall result in immediate carbon dioxide discharge bypassing the time delay relay and shall cause activation of all evacuation and discharge alarms.

2.8   CONTROL PANEL

    A.   Control panel shall be 120/208 volt, single phase, 3-wire, pre-wired in a metal cabinet, with all required relays and contacts for proper system operation, indicating lamps, nameplates, switches, terminals, supervisory circuits, battery power supply, transformer, and rectifier.

    B.   Control panel shall be of the cross-zoned type and shall be UL listed as a unit.

    C.   Each control panel shall be capable of serving all zones protected by the given carbon dioxide system plus one future zone, unless otherwise specified.

    D.   Provide local alarm indicating devices for each control panel: a 4" trouble alarm bell and an 8" fire alarm bell.

    E.   Battery Power Supply:
        1.   Uninterruptible power supply with rectifier, charger, DC bus with a battery and invertor assembly.
        2.   Battery shall be nickel cadmium, sized for 24-hour supervision and capable of carbon dioxide discharge and operation of all alarms for 15 minutes at the end of the 24-hour standby period.
        3.   Switching to the battery power in case of loss of the primary power supply shall be automatic.

2.9   ABORT SWITCHES

    A.   Abort switches may be installed in areas with 24 hour occupancy and sustained personnel training.

B. Abort switches shall be provided to manually prevent the operation of the halon system. The switches
shall be of the deadman type, requiring constant
manual pressure.

C. Abort switches shall not be used for carbon dioxide
systems protecting under floor space only.

2.10   AUDIO/VISUAL ALARMS
A. Furnish and install inside and outside of each protected area combination bell and light assemblies to
signal the first detection alarm.
1. Bell shall be of the vibrating type, 24 vdc, with
minimum sound output of 90 dB at 10 ft.
2. Light shall be 24 vdc polarized lamp unit with
high intensity light flashing at the rate of 1 flash
per second.
3. A special sign "FIRST DETECTION ALARM"
shall be installed under each assembly.

B. Furnish and install inside and outside of each protected
area a combination horn and light assembly to signal
the area evacuation upon receiving the second detection signal.
1. Horn shall be 24 vdc and shall produce steady
tone signal.
2. Light shall be 24 vdc non-polarized lamp unit
with high intensity light flashing at the rate of 1
flash per second.
3. Provide at each assembly a sign stating "WHEN
THIS ALARM IS ON, EVACUATE THE
AREA IMMEDIATELY, CARBON DIOXIDE
DISCHARGE IMMINENT."

C. Provide at each entrance to the protected area a warning light stating "DO NOT ENTER – CARBON DI-
OXIDE DISCHARGED." The light shall be turned
on automatically upon carbon dioxide discharge.

2.11   DETECTORS

A.   Detectors shall be of the ionization smoke detection type, 24 vdc, dual chamber mounted, factory calibrated, and adjusted to sensitivity for Underwriters' Laboratories Standards.

B .   Detectors shall be self-compensated for ambient temperature, humidity, and atmospheric pressure and shall not require field adjustment to compensate for above.

C .   Detectors shall have integral visual alarm indication with 300 degree field of view and shall have provisions to deter unauthorized removal.

D.   Detectors shall be fully compatible with the carbon dioxide control panel.

E.   Detectors shall be cross-zoned, i.e., located on two independent electrical circuits. Spacing and location of detectors in each zone shall comply with NFPA 72.

2.12   BREATHING APPARATUS
Provide self-contained breathing apparatus with a 30 minute air supply. They shall be located in cabinets where shown on the contract drawings.

3   CONSTRUCTION METHODS

3.1   APPLICATION
A. Installation:

   1.   Materials and equipment shall be installed in accordance with the manufacturer's recommendations and NFPA 12.

   2.   Special care shall be taken during the piping installation to avoid possible restrictions due to foreign matter or faulty fabrication.

   3.   The piping shall be securely supported with due allowance for agent thrust forces and thermal expansion and contraction.

   4.   In systems where valve arrangement introduces sections of closed piping, such sections shall be

equipped with pressure relief devices operating at between 2400 and 3000 psi.

B. Inspection and Test:

1.   Entire system performance shall be tested in accordance with requirements of all governing authorities.

2.   Upon completion of the installation, the system shall be thoroughly tested for correct operation, including testing of detection devices, all mechanical and electrical equipment, and careful inspection of all piping, hoses, and nozzles.

3.   Full carbon dioxide discharge test shall not be made. An inert gas shall be used instead of CO_2 for testing.

4.   All carbon dioxide used for initial filling, installation, and puff test purposes shall be furnished as part of the contract.

Chapter 20: Sample Specification for a Clean-Gas Fire-Protection System

1 **GENERAL**

1.1 DESCRIPTION

A. The work of this section consists of furnishing, installing, and testing a complete, automatic, clean-gas fire-extinguishing system, including storage cylinders, piping, discharge nozzles, automatic controls, ionization detectors, time-delay relay units, door and damper trips, manual pull boxes, evacuation horns, lights, alarm bell, main-to-reserve switch signs, and control panel.

The clean-gas agent work consists of the following:

1. The procurement and installation of a new, clean-gas fire-suppression system.
2. The fire-suppression system shall be of the engineered fixed-nozzle type with all pertinent components to be provided by the contractor.
3. Agent's storage cylinders shall be free-standing cylinders with wall-mounted retaining brackets. A common manifold shall be employed. The cylinders shall be installed in the new, smaller room labeled "Fire-Suppression System Room."
4. One cylinder shall be designated as the pilot cylinder and shall employ the restorable electric

actuator, mechanical manual actuator, or both. All remaining cylinders shall be pneumatically operated from the agent.

5. Manifolded cylinders shall employ a flexible discharge hose to facilitate installation and system maintenance. Each cylinder on a manifold shall also include an agent check valve installed to the manifold inlet.
6. Each cylinder shall have a pressure gauge to determine its readiness for operation.
7. Cylinders shall be installed one deep only.
8. The work also includes the furnishing and installation of an electrical/alarm panel specific for this system.

1.2   FLOW CALCULATIONS

A. Computerized verification of flow calculations shall be submitted for the gas fire-suppression system and include the following data as a minimum:
1. Quantity of agent per nozzle.
2. Orifice union/nipple and nozzle orifice diameters.
3. Pressure at nozzle.
4. Nozzle body nominal pipe size.
5. Number and size of cylinders.
6. Total agent quantity.
7. Pipe size and schedule per pipe section.
8. Number, size, and type of fitting per pipe section.
9. Actual and equivalent lengths per pipe section.
10. Discharge time.

1.3   APPLICABLE PUBLICATIONS

The publications listed below form a part of this specification to the extent referenced and latest edition shall be used as applicable. The publications are referred to in the text by basic designation only. In case of conflict between provisions of codes, laws, ordinances, and these specifications, including the contract drawings, the most stringent requirements will apply.

A.   Manufacturers Standardization Society (MSS) of the Valve and Fittings Industry Standards:
Pipe Hangers and Supports Materials, Design and Manufacture (SP58)
Pipe Hangers and Supports Selection and Application (SP 69)

B.   National Fire Protection Association (NFPA) Publications:
2001 Standard on Clean-Gas Fire Extinguishing Systems
National Electrical Code (70)
National Fire Alarm Code (72)

C.   Underwriters' Laboratories, Inc. (UL) Publications:
Fire Protection Equipment Directory
Electrical Construction Materials Directory

D.   American Society of Mechanical Engineers (ASME):
Section VIII – Pressure Vessels

E.   State Building Code

1.4   GENERAL REQUIREMENTS

A.   Standard Products:
Material and equipment shall be the standard products of a manufacturer regularly engaged in their manufacture for a minimum period of five years. All material and equipment shall be new.

B.   Nameplates:
Each major item of equipment shall have the manufacturer's name, address, type or style, model or serial number, and catalog on a plate secured to the item of equipment.

C.   Verification of Dimensions:
The Contractor shall become familiar with all details of the work, verify all dimensions in the field, and report any discrepancy before performing the work.

D.   Welding

Welding shall be in accordance with the qualified procedures as specified in Section IX of the *Pressure Vessel Code*, ASME, and in accordance with NFPA 13. Only shop welding with appropriate welding fittings shall be used.

1.5   SUBMITTALS
        A.   Shop Drawings:
             Shop drawings shall consist of illustrations, schedules, performance charts, instructions, brochures, electrical riser diagrams, elementary diagrams, wiring diagrams, schematics and complete shop drawings of the system and any other information required to illustrate that the system has been coordinated and will function as a unit. The Contractor's submittal must be signed by a registered Fire Protection Engineer regularly engaged in design of clean-gas systems or a registered Professional Engineer with a fire protection background and who has had at least two years experience in fire protection design of clean-gas systems.
        B.   Test Procedures:
             The Contractor shall furnish detailed test procedures for testing the fire-protection system.
        C.   Spare Parts Data:
             After approval of the shop drawings, the Contractor shall furnish spare parts data for each different item of material and equipment specified. The data shall include a complete list of parts and supplies with current unit prices and sources of supply and a list of the parts recommended by the manufacturer.
        D.   Operating and Maintenance Manual:
             The Contractor shall furnish complete operating instructions outlining the step-by-step procedures required for system start-up, operation, and shutdown.
        E.   Performance Test Reports:
             Upon completion and testing of the installed system, test reports shall be submitted showing all field tests

performed to prove compliance with the specified performance criteria.

1.6   SYSTEM OPERATION
The clean-gas fire-extinguishing system shall be a total flooding type as defined in NFPA 12A. Storage capacity shall provide a 100% reserve. The system shall be controlled by cross-zoned or matrixed smoke detectors and may be of the central storage or modular type.

1.7   PRODUCT DELIVERY, STORAGE, AND HANDLING
A.   Materials shall be delivered in the manufacturer's original, unopened protective packages.
B.   Materials shall be stored in their original protective packaging and protected against soiling, physical damage, or wetting before and during installation.
C.   Equipment and exposed finishes shall be protected against damage and stains during transportation and erection.
D.   Unloading of all materials shall be done in a manner to prevent misalignment or damage.

2   **MATERIALS FOR CLEAN AGENT**

2.1   PIPE MATERIAL
A.   System piping shall be of noncombustible materials having physical and chemical characteristics such that its integrity under stress can be predicted with reliability.
B.   Piping materials shall be black steel pipe conforming to ASTM A-53A ERW or ASTM A-106A seamless.
C.   Under no conditions shall ordinary cast-iron pipe, steel pipe conforming to ASTM A-120 or ASTM A-53/A-120 be used.
D.   Piping joints shall be suitable for the design conditions and shall be selected with consideration of joint tightness and mechanical strength.

E. As a minimum, fittings beyond the orifice union/ nipple shall be black 300-lb class conforming to ANSI-B-16.3. Ordinary cast-iron fittings are not acceptable. Distribution piping downstream of the orifice union must be Schedule 40.

F. The system manifold up to the orifice union/nipple must be constructed of Schedule 80 piping and 2000 lb or 3000 lb forged steel fittings.

G. All piping shall comply with NFPA 2001.

H. Piping shall be installed in accordance with good commercial practice to the appropriate codes, securely supported with UL listed hangers and arranged with close attention to the design layout since deviations may alter the design flow performance as calculated.

I. Piping shall be bracketed within 12" of all discharge nozzles.

J. All piping shall be reamed, blown clear, and swabbed with appropriate solvent to remove mill varnish and cutting oils before assembly.

K. Multi-outlet fittings, other than tees, shall not be permitted.

L. Assembly of all joints shall conform to the appropriate standards. On threaded pipe joints, TEFLON® tape shall be applied to the male threads only.

2.2 STORAGE CYLINDERS

A. Cylinder assemblies shall be of steel construction with a standard, red, epoxy paint finish. Each cylinder shall be equipped with a pressure-seat type valve and gauge. The cylinder shall utilize Ansul CV-98 forged brass valve assemblies, or equal, providing a leak-tight seal at the valve-to-cylinder connection. Each valve shall include a safety pressure-relief device, which provides relief at 3000–3360 psi per CGA test methods.

B. Initial filling and recharge shall be performed in accordance with the manufacturer's established procedures and shall not require replacement components for normal service.

2.3   CYLINDER BRACKET

    A.   Each cylinder assembly shall be furnished with a welded steel bracket. The bracket shall hold the cylinders in a saddle with a front-securing device. The brackets shall be modular in design to allow added bracketing.

    B.   Brackets shall be UL listed or FM approved for use with a clean-gas agent.

2.4   VALVE ACTUATORS

    A.   Electric valve actuators shall be of brass construction and stackable design with swivel connections to allow removal or maintenance or testing.

    B.   Actuation devices shall be UL listed and/or FM approved for use with clean-gas fire-suppression systems.

2.5   DISCHARGE HOSE/CHECK VALVE

    A.   When manifolding, all cylinder assemblies shall include a flexible discharge hose and check valve for connection to the manifold inlet.

    B.   All hose/check valves shall be UL listed and/or FM approved.

2.6   DISCHARGE NOZZLES

    A.   Discharge nozzles shall be of two-piece construction and sized to provide flow rates in accordance with system design calculations.

    B.   A nozzle inlet orifice plate shall be included. The orifice size shall be determined by a computerized UL listed flow-calculation program.

    C.   Orifice(s) shall be machined in the nozzle body to provide a horizontal discharge pattern based on the approved coverage arrangements.

    D.   Nozzles shall be permanently marked with the manufacturer's part number and threaded directly to the discharge piping without the use of special adapters.

E.   Nozzles shall be UL listed.

2.7   ORIFICE UNION/NIPPLE ASSEMBLIES
A.   An orifice union shall be included in the manifold to reduce pressure in the downstream pipe network. A separate orifice union shall be included for the extended discharge nozzles.
B.   Orifice union assemblies shall be rated at 2000-lb class minimum.
C.   Orifice union assemblies shall be permanently marked with the manufacturer's orifice code. The orifice union/nipple shall be threaded directly to the manifold piping without the use of special adapters.
D.   Orifice union assemblies shall be UL listed and/or FM approved.

2.8   ABORT SWITCH AND START SWITCH
A.   A constant, manual-pressure type push button to abort system operation shall be installed close to the room exit door. The push button shall be well marked and used only in case of a "basket fire" by the personnel working in the Protected Room.
B.   If smoke develops within the room, the alarm sounds, and people have evacuated the room, then the clean-gas agent may be released earlier by use of a manual start switch.

2.9   CONTROL SYSTEM AND COMPONENTS—DESCRIPTION AND OPERATION
A.   Control Panel
1.   The control panel shall communicate with and control the following types of equipment used to make up the system: smoke detectors, manual release/abort stations, alarm notification appliances, releasing components, and other system-controlled devices.
2.   Power supply—Input power shall be 120 V AC, 60 Hz. The power supply shall provide an inte-

gral battery charger for use with batteries up to 12 AH.

3. Batteries—Batteries shall be 2–12 V, Gell-Cell type providing 24 V DC emergency power supply. Batteries shall have sufficient capacity to power the fire-alarm system for not less than 24 hours in standby, plus 5 minutes of alarm upon a normal AC power failure. The batteries are to be completely maintenance free. No liquids are required. Fluid level checks, refilling, spills, and leakage shall not be accepted.

B. System Operation

1. Zone Status LEDs—The alarm, supervisory or trouble LED(s) shall flash until event(s) has been acknowledged. Any subsequent new alarm, supervisory, or trouble condition will re-sound all indications and flash new events.

2. Supervisory—A short circuit on this zone shall cause the supervisory LED to flash. The tone silence switch shall silence the piezo, causing the supervisory LED to illuminate steady. An open circuit shall report as a zone trouble.

3. Zone Disable—Disable/enable of any initiating circuit shall be accomplished using a special sequence of operation of the four control switches. If a zone has been disabled, an alarm shall activate the red zone LED but not the piezo or any output circuit.

4. Last-Event Recall—Last-event recall shall allow the user to display the previous panel status. Last event recall may be used to diagnose intermittent trouble conditions.

C. Counting Zone Smoke Detection—The system shall be automatically actuated by counting zone detection circuits. Smoke detectors shall be ionization detectors or photoelectric with compatibility listings for use with the control unit. Smoke detectors shall be installed at no more than 250 sq ft of coverage per detector. Detector spacing shall be in accordance with NFPA 72. The detectors shall be alternated throughout the protected area.

2.10   SEQUENCE OF OPERATION*
A.   Activation of any single detector in the detection zone shall:
1.   Cause a first-stage alarm.
2.   Energize a lamp on the activated detector and control panel.
3.   Transmit a trouble signal to the building alarm panel.
B.   Activation of a second detector in the zone shall:
1.   Transmit an alarm signal to the building alarm panel. The building alarm panel will initiate a sequence that will shut down the HVAC system (RT-2).
2.   Cause a second-stage (predischarge) alarm to operate.
3.   Operate auxiliary contact(s) for air-conditioning shutdown and automatic damper closure.
4.   Initiate a programmable time delay for the clean-agent release.
C.   Upon completion of the time delay, the clean-agent gas system shall:
1.   Cause a discharge alarm to be activated.
2.   Activate visual alarms (strobe) at protected area entrance.
3.   Energize control solenoid for the clean-agent gas cylinders, releasing gaseous agent into the protected area.

*This may be modified in accordance with local Fire Department input.

3   CONSTRUCTION METHODS

3.1   SYSTEM CHECKOUT AND TESTING

    A.   The completed installation shall be inspected by factory authorized and trained personnel. The inspection shall include a full operational test of all components per the equipment manufacturer's recommendation.

    B.   Inspection shall be performed in the presence of the owner's representative, the engineer's representative, and the Fire Department's representative.

    C.   All mechanical and electrical components shall be tested according to the manufacturer's recommended procedure to verify system integrity.

    D.   Inspection shall include a complete checkout of the detection/control system and certification of cylinder pressure. A written report shall be filed with the owner.

    E.   As-built drawings shall be provided by the contractor (2 copies) indicating the installation details. All routing of piping, electrical conduit, and accessories shall be noted.

3.2   WARRANTY

    A.   All system components furnished under this contract shall be guaranteed against defect in design, material, and workmanship for the full warranty time which is standard with the manufacturer and/or supplier but not less than one (1) year from the date of system acceptance. In addition, the installing contractor must guarantee the system against false actuation or leakage due to faulty equipment, design, or workmanship for a period of one (1) year from final acceptance. In the event of agent leakage or system discharge from any of the above conditions, the installing contractor shall completely recharge and recondition the system at no cost to the owner.

Appendix A: Alphabetical Listing of NFPA Standards

This appendix contains the complete listing of 1992 NFPA standards. For the latest list, check latest NFPA standards.

51A	Acetylene Cylinder Charging Plants—1989	Vol 2
1904	Aerial Ladder and Elevating Platform Fire Apparatus—1991	Vol 8
30B	Aerosol Products, Manufacture and Storage—1990	Vol 1
61D	Agricultural Commodities, Milling of—1989	Vol 2
90A	Air Conditioning and Ventilating Systems—1989	Vol 4
423	Aircraft Engine Test Facilities—1989	Vol 7
422M	Aircraft Fire Investigator's Manual—1989	Vol 10
415	Aircraft Fueling Ramp Drainage—1987	Vol 7
407	Aircraft Fuel Servicing—1990	Vol 7
408	Aircraft Hand Fire Extinguishers—1989	Vol 7
409	Aircraft Hangars—1990	Vol 7
417	Aircraft Loading Walkways—1990	Vol 7
410	Aircraft Maintenance—1989	Vol 7
403	Aircraft Rescue, Fire Fighting at Airports—1988	Vol 7
412	Aircraft Rescue, Evaluating Foam Fire Fighting Equipment—1987	Vol 7
402M	Aircraft Rescue and Fire Fighting Operations—1991	Vol 10
414	Aircraft Rescue and Fire Fighting Vehicles—1990	Vol 7
424M	Airport/Community Emergency Planning—1991	Vol 10
1003	Airport Fire Fighter Oualifications—1987	Vol 8
416	Airport Terminal Buildings—1987	Vol 7
419	Airport Water Supply Systems—1987	Vol 10
101M	Alternative Approaches to Life Safety—1992	Vol 9
651	Aluminum or Magnesium Powder—1987	Vol 7
65	Aluminum Processing, Finishing—1987	Vol 2
490	Ammonium Nitrate, Storage of—1986	Vol 7
232AM	Archives and Records Centers—1991	Vol 10
30A	Automotive and Marine Service Station Code—1990	Vol 1
231E	Baled Cotton Storage—1989	Vol 9

85A	Boiler-Furnaces, Oil-and Gas-Fired Single Burner—1987	Vol 4
251	Building Construction and Materials, Fire Tests of—1990	Vol 6
220	Building Construction, Standard Types of—1992	Vol 6
1402	Building Fire Service Training Centers—1992	Vol 11
703	Building Materials, Fire Retardant Coatings—1992	Vol 7
259	Building Materials, Test Method for Potential Heat—1997	Vol 6
255	Building Materials, Test of Surface Burning Characteristics of—1990	Vol 6
12	Carbon Dioxide Extinguishing Systems—1989	Vol 1
40	Cellulose Nitrate Motion Picture Film—1988	Vol 2
654	Chemical, Dye, Pharmaceutical, and Plastic Industries Prevention of Fire and Dust Explosions in—1988	Vol 7
491M	Chemical Reactions, Hazardous—1991	Vol 10
49	Chemicals Data, Hazardous—1991	Vol 9
97M	Chimneys, Vents, Heat Producing Appliances, Terms Relating to—1988	Vol 9
211	Chimneys, Vents, Fireplaces, and Solid Fuel Burning Appliances—1988	Vol 6
260	Cigarette Ignition Resistance, Components of Furniture—1989	Vol 6
261	Cigarette Ignition Resistance, Upholstered Furniture Material Assemblies—1989	Vol 6
318	Cleanrooms, Protection of—1992	Vol 6
16A	Closed-Head Foam-Water Sprinkler Systems—1988	Vol 9
1971	Clothing, Protective for Structural Fire Fighting—1991	Vol 8
120	Coal Preparation Plants—1988	Vol 6
901	Coding, Uniform for Fire Protection—1990	Vol 11
37	Combustion Engines and Gas Turbines—1990	Vol 2
85H	Combustion Hazards in Atmospheric Fluidized Bed Combustion System Boilers—1989	Vol 4
1221	Communication Systems, Public Fire Service—1991	Vol 8
473	Competencies for EMS Personnel—1992	Vol 7
52	Compressed Natural Gas (CNG) Vehicular Fuel Systems—1988	Vol 2
75	Computer/Data Processing Equipment—1989	Vol 3
241	Construction, Alteration, and Demolition Operations—1989	Vol 6
96	Cooking Equipment, Vapor Removal—1991	Vol 4
214	Cooking Towers, Water—1988	Vol 6
51B	Cutting and Welding Processes—1989	Vol 2
124	Diesel Fuel and Diesel Equipment in Underground Mines—1988	Vol 6
34	Dipping and Coating Processes Using Flammable or Combustible Liquids—1989	Vol 2

252	Door Assemblies, Fire Tests of—1990	Vol 6
17	Dry Chemical Extinguishing Systems—1990	Vol 1
32	Drycleaning Plants—1990	Vol 2
1452	Dwelling Fire Safety Surveys—1988	Vol 11
70	Electrical Code, National—1990	Vol 3
497M	Electrical Equipment in Haz. (Class.) Locations, Gases, Vapors, Dusts—1991	Vol 11
70B	Electrical Equipment Maintenance—1990	Vol 9
496	Electrical Equipment, Purged and Pressurized Enclosures for—1989	Vol 7
907M	Electrical Fires, Investigation of—1968	Vol 11
497A	Electrical Installations, Classification of Class I Hazardous Locations—1992	Vol 11
497B	Electrical Installations in Chemical Process Areas, Classification of Class II Hazardous (Classified) Locations—1991	Vol 11
70E	Electrical Safety Requirements for Employee Workplaces—1988	Vol 3
79	Electrical Standard for Industrial Machinery—1991	Vol 3
91	Exhaust Systems for Air Conveying of Materials—1992	Vol 4
69	Explosion Prevention Systems—1992	Vol 2
495	Explosive Materials Code—1990	Vol 7
498	Explosives Motor Vehicle Terminals—1990	Vol 7
80A	Exposure Fires, Protection from—1987	Vol 9
61C	Feed Mills, Fire and Dust Explosions in—1989	Vol 2
921	Fire and Explosion Investigations Guide—1992	Vol 11
262	Fire and Smoke Characteristics of Electrical Wire and Cables, Method of Test for—1990	Vol 6
1901	Fire Apparatus, Pumper—1991	Vol 8
1002	Fire Apparatus Driver/Operator Oualifications—1988	Vol 8
1410	Fire Attack, Initial—1988	Vol 8
1914	Fire Department Aerial Devices, Testing—1991	Vol 8
1931	Fire Department Ground Ladders—1989	Vol 8
1932	Fire Department Ground Ladders, Use, Maintenance, and Service Testing—1989	Vol 8
1561	Fire Department Incident Management System—1990	Vol 8
1581	Fire Department Infection Control Program—1991	Vol 8
1500	Fire Department Occupational Safety and Health Program—1987	Vol 8
13E	Fire Department Operations in Protected Properties—1989	Vol 9
1501	Fire Department Safety Officer—1987	Vol 8
1404	Fire Department Self-Contained Breathing Apparatus Program—1989	Vol 8
72E	Fire Detectors, Automatic—1990	Vol 3
80	Fire Doors and Windows—1990	Vol 4
10	Fire Extinguishers, Portable—1990	Vol 1

10L	Fire Extinguishers, Port. Enabling Act—1990	Vol 9
1001	Fire Fighter Professional Qualifications—1987	Vol 8
704	Fire Hazards of Materials, Identification—1990	Vol 7
1961	Fire Hose—1992	Vol 8
1962	Fire Hose, Care, Use, and Service Testing—1988	Vol 8
1963	Fire Hose Connections—1985	Vol 8
906M	Fire Incident Field Notes—1988	Vol 11
1031	Fire Inspector, Professional Qualifications—1997	Vol 8
1033	Fire Investigator Professional Qualifications—1987	Vol 8
1021	Fire Officer Professional Qualifications—1987	Vol 8
1	Fire Prevention Code—1987	Vol 1
20	Fire Pumps, Centrifugal—1990	Vol 1
902M	Fire Reporting Field Incident Manual—1990	Vol 11
170	Fire Safety Symbols—1991	Vol 6
1041	Fire Service Instructor Qualifications—1987	Vol 8
1983	Fire Service Life Safety Rope—1990	Vol 8
1201	Fire Services for the Public—1989	Vol 11
550	Fire Safety Concepts Tree—1986	Vol 11
74	Fire Warning Equipment, Household—1989	Vol 3
1124	Fireworks, Manufacture, Transportation, and Storage of—1988	Vol 8
1123	Fireworks, Outdoor Display of—1990	Vol 8
321	Flammable and Combustible Liquids, Classification—1991	Vol 6
30	Flammable and Combustible Liquids Code—1990	Vol 1
395	Flammable and Combustible Liquids, Farm Storage of—1988	Vol 7
386	Flammable and Combustible Liquids, Portable Shipping Tanks—1990	Vol 7
385	Flammable and Combustible Liquids, Tank Vehicles for—1990	Vol 6
329	Flammable and Combustible Liquids, Underground Leakage of—1987	Vol 10
122	Flammable and Combustible Liquids, Underground Metal and Nonmetal Mines (Other than Coal)—1990	Vol 6
328	Flammable Liquids and Gases in Manholes, Sewers—1987	Vol 10
325M	Flammable Liquids, Gases, Volatile Solids, Fire Hazard Properties of—1991	Vol 10
253	Floor Covering Systems, Critical Radiant Flux Test for—1990	Vol 6
11C	Foam Apparatus, Mobile—1990	Vol 1
298	Foam Chemicals for Wildland Fire Control—1989	Vol 6
11	Foam Extinguishing Systems, Low Expansion and Combined Agent—1988	Vol 1
11A	Foam Systems, Medium- & High-Expansion—1988	Vol 1
16	Foam-Water Sprinkler and Spray Systems—1991	Vol 1

1974	Footwear, Protective for Structural Fire Fighting—1987	Vol 8
46	Forest Products, Storage of—1990	Vol 9
850	Fossil Fueled Steam and Combustion Turbine Electric Generating Plants—1990	Vol 11
54	Fuel Gas Code, National—1988	Vol 2
81	Fur Storage, Fumigation and Cleaning—1986	Vol 4
85C	Furnace Explosions/Implosions in Multiple Burner Boiler-Furnaces—1991	Vol 4
86C	Furnaces, Industrial, Special Atmosphere—1991	Vol 4
86D	Furnaces, Industrial, Vacuum Atmosphere—1990	Vol 4
88B	Garages, Repair—1991	Vol 4
51	Gas Systems, Oxygen-Fuel, Welding, Cutting—1992	Vol 2
1973	Gloves for Structural Fire Fighting—1988	Vol 8
61B	Grain Elevators and Bulk Handling—1989	Vol 2
601	Guard Service in Fire Loss Prevention—1992	Vol 7
12B	Halon 1211 Fire Extinguishing Systems—1990	Vol 1
12A	Halon 1301 Fire Extinguishing Systems—1989	Vol 1
472	Hazardous Materials Incidents Responders, Professional Competence—1989	Vol 7
99	Health Care Facilities—1990	Vol 5
263	Heat and Visible Smoke Release Rates, Test for—1986	Vol 6
418	Heliport, Rooftop Construction and Protection—1990	Vol 7
1972	Helmets, Structural Fire Fighting—1987	Vol 8
502	Highways, Tunnels, Bridges—1987	Vol 11
913	Historic Structures and Sites—1987	Vol 11
291	Hydrants, Testing and Marking—1988	Vol 10
851	Hydroelectric Generating Plants—1987	Vol 11
50A	Hydrogen Systems, Gaseous, at Consumer Sites—1989	Vol 2
99B	Hypobaric Facilities—1990	Vol 5
904	Incident Follow-up Report Guide—1992	Vol 11
82	Incinerators, Waste and Linen Handling Systems and Equipment—1990	Vol 4
600	Industrial Fire Brigades—1992	Vol 7
505	Industrial Trucks, Powered—1987	Vol 7
1902	Initial Attack Fire Apparatus—1991	Vol 8
45	Laboratories Using Chemicals—1991	Vol 2
1405	Land-Based Fire Fighters Who Respond to Marine Vessel Fires—1990	Vol 11
910	Libraries and Library Collections—1991	Vol 11
101	Life Safety Code—1991	Vol 5
78	Lightning Protection Code—1989	Vol 3
50B	Liquefied Hydrogen Systems at Consumer Sites—1989	Vol 2
59A	Liquefied Natural Gas, Storage and Handling—1990	Vol 2
59	Liquefied Petroleum Gases at Utility Gas Plants—1992	Vol 2
58	Liquefied Petroleum Gases, Storage and Handling—1992	Vol 2

1992	Liquid Splash-Protective Suits for Hazardous	
	Chemical Emergencies—1990	Vol 8
1403	Live Fire Training Evolutions in Structures—1992	Vol 8
480	Magnesium, Storage, Handling—1987	Vol 7
501A	Manufactured Home Installations, Sites, and	
	Communities—1987	Vol 7
303	Marinas and Boatyards—1990	Vol 6
307	Marine Terminals, Piers and Wharves—1990	Vol 6
1903	Mobile Water Supply Fire Apparatus—1991	Vol 8
1125	Model Rocket Motors—1988	Vol 8
302	Motor Craft, Pleasure and Commercial—1989	Vol 6
911	Museums and Museum Collections—1991	Vol 11
72G	Notification Appliances for Protective Signaling	
	Systems—1989	Vol 9
803	Nuclear Power Plants, Light Water—1988	Vol 7
802	Nuclear Research Reactors—1988	Vol 11
31	Oil Burning Equipment, Installation of—1987	Vol 2
1981	Open-Circuit Self-Contained Breathing	
	Apparatus—1987	Vol 8
35	Organic Coatings, Manufacture of—1987	Vol 2
43B	Organic Peroxide Formulations—1986	Vol 2
86	Ovens and Furnaces, Design, Location,	
	Equipment—1990	Vol 4
43C	Oxidizing Materials, Gaseous, Storage—1986	Vol 2
43A	Oxidizers, Liquid and Solid Storage of—1990	Vol 2
50	Oxygen Systems, Bulk, at Consumer Sites—1990	Vol 2
53M	Oxygen-Enriched Atmospheres, Fire Hazards in—1990	Vol 9
88A	Parking Structures—1991	Vol 4
1982	Personal Alert Safety Systems (PASS) for Fire	
	Fighters—1988	Vol 8
43D	Pesticides in Portable Containers—1986	Vol 2
912	Places of Worship—1987	Vol 11
1141	Planned Building Groups—1990	Vol 8
650	Pneumatic Conveying Systems—1990	Vol 7
110	Power Systems, Emergency and Standby—1988	Vol 5
24	Private Fire Service Mains—1992	Vol 1
903	Property Survey Guide—1992	Vol 11
72	Protective Signaling Systems—1990	Vol 3
1035	Public Fire Educator Professional Qualifications—1987	Vol 8
85F	Pulverized Fuel Systems—1988	Vol 4
1921	Pumping Units, Portable—1987	Vol 8
1911	Pumps on Fire Department Apparatus, Service	
	Tests of—1991	Vol 8
40E	Pyroxylin Plastic—1986	Vol 2
150	Racetrack Stables—1991	Vol 6
231C	Rack Storage of Materials—1991	Vol 6

801	Radioactive Materials, Facilities Handling—1991	Vol 11
232	Records. Protection of—1991	Vol 6
501D	Recreational Vehicle Parks—1990	Vol 7
501C	Recreational Vehicles—1990	Vol 7
914	Rehabilitation and Adaptive Reuse of Historic Structures—1989	Vol 11
471	Responding to Hazardous Materials Incidents—1989	Vol 10
1122	Rockets, Code for Unmanned—1987	Vol 8
231F	Roll Paper Storage—1987	Vol 6
203	Roof Coverings—1992	Vol 9
256	Roof Coverings, Fire Tests of—1987	Vol 6
231D	Rubber Tires, Storage of—1989	Vol 6
121	Self-Propelled and Mobile Surface Mining Equipment —1990	Vol 6
71	Signaling Systems, Central Station Service—1989	Vol 3
72H	Signaling Systems, Testing—1988	Vol 9
105	Smoke-Control Door Assemblies—1989	Vol 9
204M	Smoke and Heat Venting—1991	Vol 9
92A	Smoke Control Systems—1988	Vol 9
258	Smoke Generated by Solid Materials, Standard Research Test for Measuring—1989	Vol 6
92B	Smoke Management Systems in Malls, Atria, Large Areas—1991	Vol 9
36	Solvent Extraction Plants—1988	Vol 2
33	Spray Application Using Flammable and Combustible Materials—1989	Vol 2
1964	Spray Nozzles (Shutoff and Tip)—1988	Vol 8
13A	Sprinkler Systems, Care, Maintenance—1987	Vol 9
13	Sprinkler Systems, Installation—1991	Vol 1
13D	Sprinkler Systems, One- and Two-Family Dwellings—1991	Vol 1
13R	Sprinkler Systems, Residential Occupancies up to and including Four Stories in Height—1991	
14	Standpipe and Hose Systems—1990	Vol 1
14A	Standpipe and Hose Systems, Inspection, Testing, and Maintenance—1989	Vol 9
61A	Starch Manufacturing and Handling—1989	Vol 2
77	Static Electricity—1988	Vol 9
1975	Station/Work Uniforms—1990	Vol 8
851	Stoker Operation—1989	Vol 9
231	Storage, General—1990	Vol 6
110A	Stored Electrical Energy Emergency and Standby Power Systems—1989	Vol 5
655	Sulfur Fires and Explosions, Prevention—1988	Vol 7
1993	Support Function Protective Garments for Hazardous Chemical Operations—1990	Vol 8
327	Tanks, Containers, Small, Cleaning—1987	Vol 6

297	Telecommunications Systems for Rural and Forestry Services—1986	Vol 10
102	Tents and Membrane Structures, Assembly Seating—1992	Vol 5
513	Terminals, Motor Freight—1990	Vol 7
701	Textiles and Films, Flame Resistant Test for—1989	Vol 7
481	Titanium, Handling and Storage—1987	Vol 7
1401	Training Reports and Records—1989	Vol 11
130	Transit Systems, Fixed Guideway—1990	Vol 6
512	Truck Fire Protection—1990	Vol 7
123	Underground Bituminous Coal Mines—1990	Vol 6
264A	Upholstered Furniture Components or Composites and Mattresses, Heat Release Rates for—1990	Vol 6
1991	Vapor-Protective Suits for Hazardous Chemical Emergencies—1990	Vol 8
68	Venting of Deflagrations—1988	Vol 9
306	Vessels, Control of Gas Hazards on—1988	Vol 6
312	Vessels, Protection During Construction, Repair and Lay-up—1990	Vol 6
90B	Warm Air Heating and Air Conditioning Systems—1989	Vol 5
820	Wastewater Treatment and Collection Facilities—1992	Vol 11
25	Water-Based Fire Protection Systems—1992	Vol 1
15	Water Spray Fixed Systems—1990	Vol 1
1231	Water Supplies, Suburban and Rural Fire Fighting—1989	Vol 8
26	Water Supplies, Valves Controlling—1988	Vol 9
22	Water Tanks for Private Fire Protection—1997	Vol 1
17A	Wet Chemical Extinguishing Systems—1990	Vol 1
18	Wetting Agents—1990	Vol 1
295	Wildfire Control—1991	Vol 6
299	Wildfire, Protection of Life and Property—1991	Vol 6
257	Window Assemblies, Fire Tests of—1990	Vol 6
664	Wood Processing and Woodworking, Dust Explosion—1987	Vol 7
482	Zirconium, Production, Processing—1987	Vol 7

Appendix B: Characteristics of Water Flow in Pipes

This appendix contains the following information:

Part One: Friction of Water in Pipes
Part Two: Friction Losses in Pipe Fittings
Part Three: Corresponding Pressure Table
Part Four: Symbols

Part One: Friction of Water in Pipes

Friction of Water
(Based on Darcy's Formula)
Copper Tubing—S.P.S. Copper and Brass Pipe
⅜ Inch

Flow U.S. gal per min	Type K tubing .402" inside dia .049" wall thk		Type L tubing .430" inside dia .035" wall thk		Type M tubing .450" inside dia .025" wall thk		Pipe .494" inside dia .0905" wall thk	
	Velocity ft /sec	Head loss ft /100 ft	Velocity ft /sec	Head loss ft /100 ft	Velocity ft /sec	Head loss ft /100 ft	Velocity ft /sec	Head loss ft /100 ft
0.2	0.51	0.66	0.44	0.48	0.40	0.39	0.34	0.26
0.4	1.01	2.15	0.88	1.57	0.81	1.27	0.67	0.82
0.6	1.52	4.29	1.33	3.12	1.21	2.52	1.00	1.63
0.8	2.02	7.02	1.77	5.11	1.61	4.12	1.34	2.66
1	2.52	10.32	2.20	7.50	2.01	6.05	1.68	3.89
1½	3.78	20.86	3.30	15.15	3.02	12.21	2.51	7.84
2	5.04	34.48	4.40	20.03	4.02	20.16	3.35	12.94
2½	6.30	51.03	5.50	37.01	5.03	29.80	4.19	19.11
3	7.55	70.38	6.60	51.02	6.04	41.07	5.02	26.32
3½	8.82	92.44	7.70	66.98	7.04	53.90	5.86	34.52
4	10.1	117.1	8.80	84.85	8.05	68.26	6.70	43.70
4½	11.4	144.4	9.90	104.6	9.05	84.11	7.53	53.82
5	12.6	174.3	11.0	126.1	10.05	101.4	8.36	64.87

Note: No allowance has been made for age, difference in diameter, or any abnormal condition of interior surface. Any factor of safety must be estimated from the local conditions and the requirements of each particular installation. It is recommended that for most commercial design purposes a safety factor of 15 to 20% be added to the values in the tables.

Reprinted with permission from Cameron Hydraulic *Data Book*.

Friction of Water
(Based on Darcy's Formula)
Copper Tubing—*S.P.S. Copper and Brass Pipe
½ Inch

Flow, U.S. gal per min	Type K tubing .527" inside dia .049" wall thk		Type L tubing .545" inside dia .040" wall thk		Type M tubing .569" inside dia .028" wall thk		*Pipe .625" inside dia .1075" wall thk	
	Velocity, ft /sec	Head loss, ft /100 ft	Velocity, ft /sec	Head loss, ft /100 ft	Velocity, ft /sec	Head loss, ft /100 ft	Velocity, ft /sec	Head loss, ft /100 ft
½	0.74	0.88	0.69	0.75	0.63	0.62	0.52	0.40
1	1.47	2.87	1.38	2.45	1.26	2.00	1.04	1.28
1½	2.20	5.77	2.06	4.93	1.90	4.02	1.57	2.58
2	2.94	9.52	2.75	8.11	2.53	6.61	2.09	4.24
2½	3.67	14.05	3.44	11.98	3.16	9.76	2.61	6.25
3	4.40	19.34	4.12	16.48	3.79	13.42	3.13	8.59
3½	5.14	25.36	4.81	21.61	4.42	17.59	3.66	11.25
4	5.87	32.09	5.50	27.33	5.05	22.25	4.18	14.22
4½	6.61	39.51	6.19	33.65	5.68	27.39	4.70	17.50
5	7.35	47.61	6.87	40.52	6.31	32.99	5.22	21.07
6	8.81	65.79	8.25	56.02	7.59	45.57	6.26	29.09
7	10.3	86.57	9.62	73.69	8.84	59.93	7.31	38.23
8	11.8	109.9	11.0	93.50	10.1	76.03	8.35	48.47
9	13.2	135.6	12.4	115.4	11.4	93.82	9.40	59.79
10	14.7	163.8	13.8	139.4	12.6	113.3	10.4	72.16

⅝ Inch

Flow, U.S. gal per min	Type K tubing .652" inside dia .049" wall thk		Type L tubing .666" inside dia .042" wall thk		Type M tubing .690" inside dia .030" wall thk		*Pipe	
	Velocity, ft /sec	Head loss, ft /100 ft	Velocity, ft /sec	Head loss, ft /100 ft	Velocity, ft /sec	Head loss, ft /100 ft	Velocity, ft /sec	Head loss, ft /100 ft
½	0.48	0.31	0.46	0.29	0.43	0.24		
1	0.96	1.05	0.92	0.95	0.86	0.76		
1½	1.44	2.11	1.38	1.91	1.29	1.53		
2	1.92	3.47	1.84	3.14	1.72	2.51		
2½	2.40	5.11	2.30	4.62	2.14	3.68		
3	2.88	7.02	2.75	6.35	2.57	5.07		
3½	3.36	9.20	3.21	8.32	3.00	6.64		
4	3.84	11.63	3.67	10.51	3.43	8.40		
4½	4.32	14.30	4.13	12.93	3.86	10.35		
5	4.80	17.22	4.59	15.56	4.29	12.49		
6	5.75	23.76	5.51	21.47	5.15	17.21		
7	6.71	31.22	6.42	28.21	6.00	22.58		
8	7.67	39.58	7.35	35.75	6.85	28.54		
9	8.64	48.81	8.25	44.09	7.71	35.35		
10	9.60	58.90	9.18	53.19	8.57	42.48		
11	10.6	69.83	10.1	63.06	9.43	50.47		
12	11.5	81.59	11.0	73.67	10.3	59.1		
13	12.5	94.18	11.9	85.03	11.2	68.8		

Note: No allowance has been made for age, difference in diameter, or any abnormal condition of interior surface. Any factor of safety must be estimated from the local conditions and the requirements of each particular installation. It is recommended that for most commercial design purposes a safety factor of 15 to 20% be added to the values in the tables.

Reprinted with permission from Cameron Hydraulic *Data Book*.

Friction of Water
(Based on Darcy's Formula)
Copper Tubing—*S.P.S. Copper and Brass Pipe
¾ Inch

Flow, U.S.	Type K tubing		Type L tubing		Type M tubing		*Pipe	
	.745" inside dia .065" wall thk		.785" inside dia .045" wall thk		.811" inside dia .032" wall thk		.822" inside dia .114" wall thk	
gal per min	Velocity, ft /sec	Head loss, ft /100 ft	Velocity, ft /sec	Head loss, ft /100 ft	Velocity, ft /sec	Head loss, ft /100 ft	Velocity, ft /sec	Head loss, ft /100 ft
1	0.74	0.56	0.66	0.44	0.62	0.38	0.60	0.35
2	1.47	1.84	1.33	1.44	1.24	1.23	1.21	1.16
3	2.21	3.73	1.99	2.91	1.86	2.49	1.81	2.34
4	2.94	6.16	2.65	4.81	2.48	4.12	2.42	3.86
5	3.67	9.12	3.31	7.11	3.10	6.09	3.02	5.71
6	4.41	12.57	3.98	9.80	3.72	8.39	3.62	7.86
7	5.14	16.51	4.64	12.86	4.34	11.01	4.23	10.32
8	5.88	20.91	5.30	16.28	4.96	13.94	4.83	13.07
9	6.61	25.77	5.96	20.06	5.59	17.17	5.44	16.10
10	7.35	31.08	6.62	24.19	6.20	20.70	6.04	19.41
11	8.09	36.83	7.29	28.66	6.82	24.52	6.64	22.99
12	8.83	43.01	7.95	33.47	7.44	28.63	7.25	26.84
13	9.56	49.62	8.61	38.61	8.06	33.02	7.85	30.96
14	10.3	56.66	9.27	44.07	8.68	37.69	8.45	35.33
15	11.0	64.11	9.94	49.86	9.30	42.64	9.05	39.97
16	11.8	71.97	10.6	55.97	9.92	47.86	9.65	44.86
17	12.5	80.24	11.25	62.39	10.55	53.35	10.25	50.00
18	13.2	88.92	11.92	69.13	11.17	59.10	10.85	55.40

1 Inch

Flow U.S.	Type K tubing		Type L tubing		Type M tubing		*Pipe	
	.995" inside dia .065" wall thk		1.025" inside dia .050" wall thk		1.055" inside dia .035" wall thk		1.062" inside dia .1265" wall thk	
gal per min	Velocity ft /sec	Head loss ft /100 ft	Velocity ft /sec	Head loss ft /100 ft	Velocity ft /sec	Head loss ft /100 ft	Velocity ft /sec	Head loss ft /100 ft
2	0.82	0.47	0.78	0.41	0.73	0.36	0.72	0.35
3	1.24	0.95	1.17	0.82	1.10	0.72	1.08	0.70
4	1.65	1.56	1.56	1.35	1.47	1.18	1.45	1.14
5	2.06	2.30	1.95	2.00	1.83	1.74	1.81	1.69
6	2.48	3.17	2.34	2.75	2.20	2.40	2.17	2.32
7	2.89	4.15	2.72	3.60	2.56	3.14	2.53	3.04
8	3.30	5.25	3.11	4.56	2.93	3.97	2.89	3.85
9	3.71	6.47	3.50	5.61	3.30	4.89	3.25	4.74
10	4.12	7.79	3.89	6.76	3.66	5.89	3.61	5.71
12	4.95	10.76	4.67	9.33	4.40	8.13	4.34	7.88
14	5.77	14.15	5.45	12.27	5.13	10.69	5.05	10.36
16	6.60	17.94	6.22	15.56	5.86	13.55	5.78	13.13
18	7.42	22.14	7.00	19.20	6.60	16.72	6.50	16.20
20	8.24	26.73	7.78	23.18	7.33	20.18	7.22	19.55
25	10.30	39.87	9.74	34.56	9.16	30.09	9.03	29.15
30	12.37	55.33	11.68	47.96	11.00	41.74	10.84	40.43
35	14.42	73.06	13.61	63.31	12.82	55.09	12.65	53.37
40	16.50	93.00	15.55	80.58	14.66	70.11	14.45	67.90
45	18.55	115.1	17.50	99.72	16.50	86.75	16.25	84.02
50	20.60	139.4	19.45	120.7	18.32	105.0	18.05	101.7

Note: No allowance has been made for age, difference in diameter, or any abnormal condition of interior surface. Any factor of safety must be estimated from the local conditions and the requirements of each particular installation. It is recommended that for most commercial design purposes a safety factor of 15 to 20% be added to the values in the tables.

Reprinted with permission from Cameron Hydraulic Data Book.

Friction of Water
(Based on Darcy's Formula)
Copper Tubing—*S.P.S. Copper and Brass Pipe
1¼ Inch

Flow, U.S.	Type K tubing 1.245" inside dia .065" wall thk		Type L tubing 1.265" inside dia .055" wall thk		Type M tubing 1.291" inside dia .042" wall thk		*Pipe 1.368" inside dia .146" wall thk	
gal per min	Velocity, ft /sec	Head loss, ft /100 ft	Velocity, ft /sec	Head loss, ft /100 ft	Velocity, ft /sec	Head loss, ft /100 ft	Velocity, ft /sec	Head loss, ft /100 ft
5	1.31	0.79	1.28	0.74	1.22	0.67	1.09	0.51
6	1.58	1.09	1.53	1.01	1.47	0.92	1.31	0.70
7	1.84	1.43	1.79	1.32	1.71	1.20	1.53	0.91
8	2.11	1.81	2.04	1.67	1.96	1.52	1.75	1.15
9	2.37	2.22	2.30	2.06	2.20	1.87	1.96	1.42
10	2.63	2.67	2.55	2.48	2.45	2.25	2.18	1.71
12	3.16	3.69	3.06	3.42	2.93	3.10	2.62	2.35
15	3.95	5.47	3.83	5.07	3.66	4.60	3.27	3.49
20	5.26	9.13	5.10	8.46	4.89	7.67	4.36	5.81
25	6.58	13.59	6.38	12.59	6.11	11.42	5.46	8.65
30	7.90	18.83	7.65	17.44	7.33	15.82	6.55	11.98
35	9.21	24.83	8.94	23.00	8.55	20.86	7.65	15.79
40	10.5	31.57	10.2	29.24	9.77	26.51	8.74	20.06
45	11.8	38.03	11.5	36.15	11.0	32.77	9.83	24.80
50	13.2	47.20	12.8	43.71	12.2	39.63	10.9	29.98
60	15.8	65.65	15.3	60.78	14.7	55.10	13.1	41.66
70	18.4	86.82	17.9	80.38	17.1	72.86	15.3	58.07
80	21.1	110.7	20.4	102.5	19.6	92.85	17.5	70.16
90	23.7	137.2	23.0	127.0	22.0	115.1	19.6	86.91
100	26.3	166.3	25.5	153.9	24.4	139.4	21.8	105.3

1½ Inch

Flow, U.S.	Type K tubing 1.481" inside dia .072" wall thk		Type L tubing 1.505" inside dia .060" wall thk		Type M tubing 1.527" inside dia .049" wall thk		*Pipe 1.600" inside dia .150" wall thk	
gal per min	Velocity, ft /sec	Head loss, ft /100 ft	Velocity, ft /sec	Head loss, ft /100 ft	Velocity, ft /sec	Head loss, ft /100 ft	Velocity, ft /sec	Head loss, ft /100 ft
8	1.49	0.79	1.44	0.73	1.40	0.68	1.27	0.55
9	1.67	0.97	1.62	0.90	1.57	0.84	1.43	0.67
10	1.86	1.17	1.80	1.08	1.75	1.01	1.59	0.81
12	2.23	1.61	2.16	1.49	2.10	1.39	1.91	1.12
15	2.79	2.39	2.70	2.21	2.63	2.07	2.39	1.65
20	3.72	3.98	3.60	3.68	3.50	3.44	3.19	2.75
25	4.65	5.91	4.51	5.48	4.38	5.11	3.98	4.09
30	5.58	8.19	5.41	7.58	5.25	7.07	4.78	5.65
35	6.51	10.79	6.31	9.99	6.13	9.31	5.58	7.45
40	7.44	13.70	7.21	12.68	7.00	11.83	6.37	9.45
45	8.37	16.93	8.11	15.67	7.88	14.61	7.16	11.68
50	9.30	20.46	9.01	18.94	8.76	17.66	7.96	14.11
60	11.2	28.42	10.8	26.30	10.5	24.53	9.56	19.59
70	13.0	37.55	12.6	34.74	12.3	32.40	11.2	25.87
80	14.9	47.82	14.4	44.24	14.0	41.25	12.8	32.93
90	16.7	59.21	16.2	54.78	15.8	51.07	14.4	40.76
100	18.6	71.70	18.0	66.34	17.5	61.84	15.9	49.34
110	20.5	85.29	19.8	78.90	19.3	73.55	17.5	56.67
120	22.3	99.95	21.6	92.46	21.0	86.18	19.1	68.74
130	24.2	115.7	23.4	107.0	22.8	99.73	20.7	79.53

Note: No allowance has been made for age, difference in diameter, or any abnormal condition of interior surface. Any factor of safety must be estimated from the local conditions and the requirements of each particular installation. It is recommended that for most commercial design purposes a safety factor of 15 to 20% be added to the values in the tables.

Reprinted with permission from Cameron Hydraulic *Data Book*.

Friction of Water
(Based on Darcy's Formula)
Copper Tubing—*S.P.S. Copper and Brass Pipe
2 Inch

Flow, U.S. gal per min	Type K tubing 1.959" inside dia .083" wall thk Velocity, ft /sec	Head loss, ft /100 ft	Type L tubing 1.985" inside dia .070" wall thk Velocity, ft /sec	Head loss, ft /100 ft	Type M tubing 2.009" inside dia .058" wall thk Velocity, ft /sec	Head loss, ft /100 ft	*Pipe 2.062" inside dia .1565" wall thk Velocity, ft /sec	Head loss, ft /100 ft
10	1.07	0.31	1.04	0.29	1.01	0.27	0.96	0.24
12	1.28	0.43	1.24	0.40	1.21	0.38	1.15	0.33
14	1.49	0.56	1.45	0.52	1.42	0.50	1.34	0.44
16	1.70	0.71	1.66	0.66	1.62	0.63	1.53	0.55
18	1.92	0.87	1.87	0.82	1.82	0.77	1.72	0.68
20	2.13	1.05	2.07	0.98	2.02	0.93	1.92	0.82
25	2.66	1.55	2.59	1.46	2.53	1.38	2.39	1.22
30	3.19	2.15	3.11	2.01	3.03	1.90	2.87	1.68
35	3.73	2.82	3.62	2.65	3.54	2.50	3.35	2.21
40	4.26	3.58	4.14	3.36	4.05	3.17	3.83	2.80
45	4.79	4.42	4.66	4.15	4.55	3.92	4.30	3.46
50	5.32	5.34	5.17	5.01	5.05	4.73	4.80	4.17
60	6.39	7.40	6.21	6.95	6.06	6.56	5.75	5.79
70	7.45	9.76	7.25	9.16	7.07	8.65	6.70	7.63
80	8.52	12.42	8.28	11.65	8.09	11.00	7.65	9.70
90	9.58	15.36	9.31	14.41	9.10	13.60	8.61	12.00
100	10.65	18.58	10.4	17.43	10.1	16.45	9.57	14.51
110	11.71	22.07	11.4	20.71	11.1	19.55	10.5	17.24
120	12.78	25.84	12.4	24.25	12.1	22.88	11.5	20.18
130	13.85	29.88	13.4	28.04	13.1	26.45	12.5	23.33
140	14.9	34.18	14.5	32.07	14.2	30.26	13.4	26.69
150	16.0	38.75	15.5	36.36	15.2	34.30	14.4	30.25
160	17.0	43.58	16.5	40.89	16.2	38.58	15.3	34.01
170	18.1	48.67	17.6	45.66	17.2	43.08	16.3	37.98
180	19.2	54.01	18.6	50.67	18.2	47.81	17.2	42.15
190	20.2	59.61	19.6	55.92	19.2	52.76	18.2	46.51
200	21.3	65.46	20.7	61.41	20.2	57.94	19.2	51.07
210	22.4	71.57	21.7	67.14	21.2	63.34	20.1	55.83
220	23.4	77.93	22.8	73.10	22.2	68.96	21.0	60.78
230	24.5	84.53	23.8	79.29	23.2	74.80	22.0	65.93
240	25.6	91.38	24.8	85.72	24.3	80.86	23.0	71.26
250	26.6	98.43	25.9	92.37	25.3	87.14	23.9	76.79
260	27.7	105.8	26.9	99.26	26.3	93.63	24.9	82.51
270	28.8	113.4	27.9	106.4	27.3	100.3	25.8	88.42
280	29.8	121.3	29.0	113.7	28.3	107.3	26.8	94.52
290	30.9	129.3	30.0	121.3	29.4	114.4	27.8	100.8
300	32.0	137.6	31.1	129.1	30.4	121.8	28.7	107.3

Note: No allowance has been made for age, difference in diameter, or any abnormal condition of interior surface. Any factor of safety must be estimated from the local conditions and the requirements of each particular installation. It is recommended that for most commercial design purposes a safety factor of 15 to 20% be added to the values in the tables.

Reprinted with permission from Cameron Hydraulic Data Book.

Friction of Water
(Based on Darcy's Formula)
Copper Tubing—*S.P.S. Copper and Brass Pipe
2½ Inch

Flow, U.S. gal per min	Type K tubing 2.435" inside dia .095" wall thk		Type L tubing 2.465" inside dia .080" wall thk		Type M tubing 2.495" inside dia .065" wall thk		*Pipe 2.500" inside dia .1875" wall thk	
	Velocity, ft /sec	Head loss, ft /100 ft	Velocity, ft /sec	Head loss, ft /100 ft	Velocity, ft /sec	Head loss, ft /100 ft	Velocity, ft /sec	Head loss, ft /100 ft
20	1.38	0.37	1.34	0.35	1.31	0.33	1.31	0.33
25	1.72	0.55	1.68	0.52	1.64	0.49	1.63	0.49
30	2.07	0.76	2.02	0.72	1.97	0.68	1.96	0.67
35	2.41	1.00	2.35	0.94	2.30	0.89	2.29	0.88
40	2.76	1.26	2.69	1.19	2.62	1.13	2.61	1.12
45	3.10	1.56	3.02	1.47	2.95	1.39	2.94	1.38
50	3.45	1.88	3.36	1.77	3.28	1.68	3.26	1.66
60	4.14	2.61	4.03	2.46	3.93	2.32	3.92	2.30
70	4.82	3.43	4.70	3.24	4.59	3.06	4.57	3.03
80	5.51	4.36	5.37	4.12	5.25	3.88	5.22	3.85
90	6.20	5.39	6.04	5.08	5.90	4.80	5.88	4.75
100	6.89	6.52	6.71	6.15	6.55	5.80	6.53	5.74
110	7.58	7.74	7.38	7.30	7.21	6.89	7.19	6.82
120	8.27	9.06	8.05	8.54	7.86	8.05	7.84	7.98
130	8.96	10.46	8.73	9.87	8.52	9.31	8.49	9.22
140	9.65	11.97	9.40	11.28	9.18	10.64	9.14	10.54
150	10.35	13.56	10.1	12.78	9.83	12.06	9.79	11.94
160	11.0	15.24	10.8	14.36	10.5	13.55	10.45	13.42
170	11.7	17.01	11.4	16.03	11.1	15.12	11.1	14.98
180	12.4	18.87	12.1	17.79	11.8	16.78	11.8	16.61
190	13.1	20.81	12.8	19.62	12.5	18.51	12.4	18.33
200	13.8	22.85	13.4	21.54	13.1	20.31	13.1	20.12
220	15.2	27.18	14.8	25.61	14.4	24.16	14.4	23.93
240	16.5	31.84	16.1	30.01	15.7	28.31	15.7	28.03
260	17.9	36.85	17.5	34.73	17.1	32.75	17.0	32.44
280	19.3	42.19	18.8	39.76	18.4	37.50	18.3	37.13
300	20.7	47.86	20.1	45.10	19.7	42.53	19.6	42.12
320	22.1	53.86	21.5	50.75	21.0	47.86	20.9	47.40
340	23.4	60.18	22.8	56.71	22.3	53.48	22.2	52.96
360	24.8	66.83	24.2	62.97	23.6	59.38	23.5	58.81
380	26.2	73.80	25.5	69.54	24.9	65.57	24.8	64.94
400	27.6	81.09	26.9	76.41	26.2	72.04	26.1	71.35
420	29.0	88.70	28.2	83.57	27.5	78.80	27.4	78.04
440	30.3	96.62	29.5	91.04	28.8	85.83	28.7	85.00
460	31.7	104.9	30.9	98.80	30.2	93.15	30.0	92.24
480	33.1	113.4	32.2	106.8	31.5	100.7	31.4	99.76
500	34.5	122.3	33.6	115.2	32.8	108.6	32.6	107.5

Note: No allowance has been made for age, difference in diameter, or any abnormal condition of interior surface. Any factor of safety must be estimated from the local conditions and the requirements of each particular installation. It is recommended that for most commercial design purposes a safety factor of 15 to 20% be added to the values in the tables.

Reprinted with permission from Cameron Hydraulic *Data Book*.

Friction of Water
(Based on Darcy's Formula)
Copper Tubing—*S.P.S. Copper and Brass Pipe
3 Inch

Flow, U.S. gal per min	Type K tubing 2.907" inside dia .109" wall thk		Type L tubing 2.945" inside dia .090" wall thk		Type M tubing 2.981" inside dia .072" wall thk		*Pipe 3.062" inside dia .219" wall thk	
	Velocity, ft /sec	Head loss, ft /100 ft	Velocity, ft /sec	Head loss, ft /100 ft	Velocity, ft /sec	Head loss, ft /100 ft	Velocity, ft /sec	Head loss, ft /100 ft
20	0.96	0.16	0.94	0.15	0.92	0.14	0.87	0.13
30	1.45	0.33	1.41	0.31	1.37	0.29	1.30	0.25
40	1.93	0.54	1.88	0.51	1.83	0.48	1.74	0.42
50	2.41	0.81	2.35	0.76	2.29	0.72	2.17	0.63
60	2.89	1.12	2.82	1.05	2.75	0.99	2.61	0.87
70	3.38	1.47	3.29	1.38	3.20	1.30	3.04	1.15
80	3.86	1.87	3.76	1.75	3.66	1.65	3.48	1.45
90	4.34	2.30	4.23	2.16	4.12	2.04	3.91	1.80
100	4.82	2.78	4.70	2.61	4.59	2.47	4.35	2.17
110	5.30	3.30	5.17	3.10	5.05	2.93	4.79	2.57
120	5.79	3.86	5.64	3.63	5.50	3.42	5.21	3.01
130	6.27	4.46	6.11	4.19	5.95	3.95	5.65	3.47
140	6.75	5.10	6.58	4.79	6.41	4.52	6.09	3.97
150	7.24	5.77	7.05	5.42	6.87	5.12	6.52	4.50
160	7.72	6.49	7.52	6.09	7.34	5.75	6.95	5.05
170	8.20	7.24	7.99	6.80	7.79	6.41	7.39	5.64
180	8.69	8.03	8.46	7.54	8.25	7.11	7.82	6.25
190	9.16	8.85	8.93	8.32	8.70	7.84	8.25	6.89
200	9.64	9.71	9.40	9.13	9.16	8.61	8.70	7.56
220	10.6	11.55	10.3	10.85	10.1	10.23	9.56	8.99
240	11.6	13.52	11.3	12.70	11.0	11.98	10.4	10.52
260	12.6	15.64	12.2	14.69	11.9	13.85	11.3	12.17
280	13.5	17.90	13.2	16.81	12.8	15.85	12.2	13.93
300	14.5	20.30	14.1	19.06	13.7	17.97	13.0	15.79
320	15.4	22.83	15.0	21.44	14.7	20.22	13.9	17.76
340	16.4	25.50	16.0	23.95	15.6	22.58	14.8	19.83
360	17.4	28.30	16.9	26.50	16.5	25.06	15.7	22.01
380	18.3	31.24	17.9	29.34	17.4	27.66	16.5	24.29
400	19.3	34.32	18.8	32.22	18.3	30.38	17.4	26.68
450	21.7	42.58	21.2	39.98	20.6	37.69	19.6	33.09
500	24.1	51.65	23.5	48.50	22.9	45.72	21.7	40.14
550	26.6	61.54	25.8	57.77	25.2	54.46	23.9	47.81
600	29.0	72.22	28.2	67.80	27.5	63.91	26.1	56.10
650	31.4	83.69	30.6	78.56	29.8	74.05	28.2	65.00
700	33.8	95.95	32.9	90.06	32.1	84.89	30.4	74.50
750	36.2	109.0	35.2	102.3	34.4	96.41	32.6	84.61
800	38.6	122.8	37.6	115.3	36.6	108.6	34.8	95.31

Note: No allowance has been made for age, difference in diameter, or any abnormal condition of interior surface. Any factor of safety must be estimated from the local conditions and the requirements of each particular installation. It is recommended that for most commercial design purposes a safety factor of 15 to 20% be added to the values in the tables.

Reprinted with permission from Cameron Hydraulic *Data Book.*

Friction of Water – New Steel Pipe
(Based on Darcy's Formula)
¼ Inch

Flow, U.S. gal per min	Standard wt steel – sch 40			Extra strong steel – sch 80		
	0.364" inside dia			0.302" inside dia		
	Velocity, ft per sec	Velocity head, ft	Head loss, ft per 100 ft	Velocity, ft per sec	Velocity head, ft	Head loss, ft per 100 ft
0.4	1.23	0.024	3.7	1.79	0.05	9.18
0.6	1.85	0.053	7.6	2.69	0.11	19.0
0.8	2.47	0.095	12.7	3.59	0.20	32.3
1.0	3.08	0.148	19.1	4.48	0.31	48.8
1.2	3.70	0.213	26.7	5.38	0.45	68.6
1.4	4.32	0.290	35.6	6.27	0.61	91.7
1.6	4.93	0.378	45.6	7.17	0.80	118.1
1.8	5.55	0.479	56.9	8.07	1.01	147.7
2.0	6.17	0.591	69.4	8.96	1.25	180.7
2.4	7.40	0.850	98.1	10.75	1.79	256
2.8	8.63	1.157	132	12.54	2.44	345

⅜ Inch

Flow, U.S. gal per min	Standard wt steel – sch 40			Extra strong steel – sch 80		
	0.493" inside dia			0.423" inside dia		
	Velocity, ft per sec	Velocity head, ft	Head loss, ft per 100 ft	Velocity, ft per sec	Velocity head, ft	Head loss, ft per 100 ft
0.5	0.84	0.011	1.26	1.14	0.02	2.63
1.0	1.68	0.044	4.26	2.28	0.08	9.05
1.5	2.52	0.099	8.85	3.43	0.18	19.0
2.0	3.36	0.176	15.0	4.57	0.32	32.4
2.5	4.20	0.274	22.7	5.71	0.51	49.3
3.0	5.04	0.395	32.0	6.85	0.73	69.6
3.5	5.88	0.538	42.7	8.00	0.99	93.3
4.0	6.72	0.702	55.0	9.14	1.30	120
5.0	8.40	1.097	84.2	11.4	2.0	185
6.0	10.08	1.58	119	13.7	2.9	263

Calculations on pages 3-12 to 3-34 are by Ingersoll-Rand Co.

Note: No allowance has been made for age, difference in diameter, or any abnormal condition of interior surface. Any factor of safety must be estimated from the local conditions and the requirements of each particular installation. It is recommended that for most commercial design purposes a safety factor of 15 to 20% be added to the values in the tables.

Reprinted with permission from Cameron Hydraulic *Data Book*.

Friction of Water – New Steel Pipe
(Based on Darcy's Formula)
½ Inch

Flow, U.S. gal per min	Standard wt steel – sch 40 .622" inside dia			Extra strong steel – sch 80 .546" inside dia			Steel – Schedule 160 .464" inside dia		
	Velocity, ft /sec	Velocity head, ft	Head loss, ft /100 ft	Velocity, ft /sec	Velocity head, ft	Head loss, ft /100 ft	Velocity, ft /sec	Velocity head, ft	Head loss, ft /100 ft
0.7	0.739	.008	0.74	.96	.01	1.39			
1.0	1.056	.017	1.86	1.37	.03	2.58	1.90	.056	1.68
1.5	1.58	.039	2.82	2.06	.07	5.34	2.85	.126	5.73
2.0	2.11	.069	4.73	2.74	.12	9.02	3.80	.224	12.0
2.5	2.64	.108	7.10	3.43	.18	13.6	4.74	.349	20.3
3.0	3.17	.156	9.94	4.11	.26	19.1	5.69	.503	30.8
3.5	3.70	.212	13.2	4.80	.36	25.5	6.64	.684	43.5
4.0	4.22	.277	17.0	5.48	.47	32.7	7.59	.894	58.2
4.5	4.75	.351	21.1	6.17	.59	40.9	8.54	1.13	75.0
5.0	5.28	.433	25.8	6.86	.73	50.0	9.49	1.40	94.0
5.5	5.81	.524	30.9	7.54	.88	59.9	10.44	1.69	115
6.0	6.34	.624	36.4	8.23	1.05	70.7	11.38	2.01	138
6.5	6.86	.732	42.4	8.91	1.23	82.4	12.33	2.36	163
7.0	7.39	.849	48.8	9.60	1.43	95.0	13.28	2.74	190
7.5	7.92	.975	55.6	10.3	1.6	109	14.23	3.14	220
8.0	8.45	1.109	63.0	11.0	1.9	123			
8.5	8.98	1.25	70.7	11.6	2.1	138			
9.0	9.50	1.40	78.9	12.3	2.4	154			
9.5	10.03	1.56	87.6	13.0	2.6	171			
10	10.56	1.73	96.6	13.7	2.9	189			

¾ Inch

Flow, U.S. gal per min	Standard wt steel – sch 40 .824" inside dia			Extra strong steel – sch 80 .742" inside dia			Steel – Schedule 160 .612" inside dia		
	Velocity, ft /sec	Velocity head, ft	Head loss, ft /100 ft	Velocity, ft /sec	Velocity head, ft	Head loss, ft /100 ft	Velocity, ft /sec	Velocity head, ft	Head loss, ft /100 ft
1.5	0.90	.013	0.72	1.11	.02	1.19	1.64	.042	3.05
2.0	1.20	.023	1.19	1.48	.03	1.99	2.18	.074	5.12
2.5	1.50	.035	1.78	1.86	.05	2.97	2.73	.115	7.70
3.0	1.81	.051	2.47	2.23	.08	4.14	3.27	.166	10.8
3.5	2.11	.069	3.26	2.60	.11	5.48	3.82	.226	14.3
4.0	2.41	.090	4.16	2.97	.14	7.01	4.36	.295	18.4
4.5	2.71	.114	5.17	3.34	.17	8.72	4.91	.374	22.9
5.0	3.01	.141	6.28	3.71	.21	10.6	5.45	.462	28.0
6	3.61	.203	8.80	4.45	.31	14.9	6.54	.665	39.5
7	4.21	.276	11.7	5.20	.42	19.9	7.64	.905	53.0
8	4.81	.360	15.1	5.94	.55	25.6	8.73	1.18	68.4
9	5.42	.456	18.8	6.68	.69	32.1	9.82	1.50	85.8
10	6.02	.563	23.0	7.42	.86	39.2	10.91	1.85	105
11	6.62	.681	27.6	8.17	1.04	47.0	12.00	2.23	126
12	7.22	.722	32.5	8.91	1.23	55.5	13.09	2.66	149
13	7.82	.951	37.9	9.63	1.44	64.8	14.18	3.13	175
14	8.42	1.103	43.7	10.4	1.7	74.7	15.27	3.62	202
16	9.63	1.44	56.4	11.9	2.2	96.7	17.45	4.73	261
18	10.8	1.82	70.8	13.4	2.8	121			
20	12.0	2.25	86.8	14.8	3.4	149			

Note: No allowance has been made for age, difference in diameter, or any abnormal condition of interior surface. Any factor of safety must be estimated from the local conditions and the requirements of each particular installation. It is recommended that for most commercial design purposes a safety factor of 15 to 20% be added to the values in the tables.

Reprinted with permission from Cameron Hydraulic Data Book.

Friction of Water – New Steel Pipe
(Based on Darcy's Formula)
1 Inch

Flow, U.S.	Standard wt steel – sch 40			Extra strong steel – sch 80			Steel – Schedule 160		
	1.049" inside dia			.957" inside dia			.815" inside dia		
gal per min	Velocity, ft /sec	Velocity head, ft	Head loss, ft /100 ft	Velocity, ft /sec	Velocity head, ft	Head loss, ft /100 ft	Velocity, ft /sec	Velocity head, ft	Head loss, ft /100 ft
2	0.74	.009	.385	.89	.01	.599	1.23	.023	1.26
3	1.11	.019	.787	1.34	.03	1.19	1.85	.053	2.60
4	1.48	.034	1.270	1.79	.05	1.99	2.46	.094	4.40
5	1.86	.054	1.90	2.23	.08	2.99	3.08	.147	6.63
6	2.23	.077	2.65	2.68	.11	4.17	3.69	.211	9.30
8	2.97	.137	4.50	3.57	.20	7.11	4.92	.376	15.9
10	3.71	.214	6.81	4.46	.31	10.8	6.15	.587	24.3
12	4.45	.308	9.58	5.36	.45	15.2	7.38	.845	34.4
14	5.20	.420	12.8	6.25	.61	20.4	8.61	1.15	46.2
16	5.94	.548	16.5	7.14	.79	26.3	9.84	1.50	59.7
18	6.68	.694	20.6	8.03	1.00	32.9	11.07	1.90	74.9
20	7.42	.857	25.2	8.92	1.24	40.3	12.30	2.35	91.8
22	8.17	1.036	30.3	9.82	1.50	48.4	13.53	2.84	110
24	8.91	1.23	36.8	10.7	1.8	57.2	14.76	3.38	131
26	9.65	1.45	41.7	11.6	2.1	66.8	15.99	3.97	153
28	10.39	1.68	48.1	12.5	2.4	77.1			
30	11.1	1.93	55.0	13.4	2.8	88.2			
35	13.0	2.62	74.1	15.6	3.8	119			
40	14.8	3.43	96.1	17.9	5.0	154			
45	16.7	4.33	121	20.1	6.3	194			

1¼ Inch

Flow, U.S.	Standard wt steel – sch 40			Extra strong steel – sch 80			Steel – Schedule 160		
	1.380" inside dia			1.278" inside dia			1.160" inside dia		
gal per min	Velocity, ft /sec	Velocity head, ft	Head loss, ft /100 ft	Velocity, ft /sec	Velocity head, ft	Head loss, ft /100 ft	Velocity, ft /sec	Velocity head, ft	Head loss, ft /100 ft
4	.858	.011	.35	1.00	.015	.51	1.21	.023	.806
5	1.073	.018	.52	1.25	.024	.75	1.52	.036	1.20
6	1.29	.026	.72	1.50	.034	1.04	1.82	.051	1.61
7	1.50	.035	.95	1.75	.048	1.33	2.13	.070	2.14
8	1.72	.046	1.20	2.00	.062	1.69	2.43	.092	2.73
10	2.15	.072	1.74	2.50	.097	2.55	3.04	.143	4.12
12	2.57	.103	2.45	3.00	.140	3.57	3.64	.206	5.78
14	3.00	.140	3.24	3.50	.190	4.75	4.25	.280	7.72
16	3.43	.183	4.15	4.00	.249	6.10	4.86	.366	9.92
18	3.86	.232	5.17	4.50	.315	7.61	5.46	.463	12.4
20	4.29	.286	6.31	5.00	.388	9.28	6.07	.572	15.1
25	5.36	.431	9.61	6.25	.607	14.2	7.59	.894	23.2
30	6.44	.644	13.6	7.50	.874	20.1	9.11	1.29	32.9
35	7.51	.876	18.2	8.75	1.19	27.0	10.63	1.75	44.2
40	8.58	1.14	23.5	10.0	1.55	34.9	12.14	2.29	57.3
50	10.7	1.79	36.2	12.5	2.43	53.7	15.18	3.58	88.3
60	12.9	2.57	51.5	15.0	3.50	76.5	18.22	5.15	126
70	15.0	3.50	69.5	17.5	4.76	103	21.25	7.01	170
80	17.2	4.53	90.2	20.0	6.21	134	24.29	9.16	221
90	19.3	5.79	114	22.5	7.86	168	27.32	11.59	279

Note: No allowance has been made for age, difference in diameter, or any abnormal condition of interior surface. Any factor of safety must be estimated from the local conditions and the requirements of each particular installation. It is recommended that for most commercial design purposes a safety factor of 15 to 20% be added to the values in the tables.

Reprinted with permission from Cameron Hydraulic *Data Book*.

Friction of Water – New Steel Pipe
(Based on Darcy's Formula)
1½ Inch

Flow, U.S. gal per min	Standard wt steel – sch 40			Extra strong steel – sch 80			Steel – Schedule 160		
	1.610" inside dia			1.500" inside dia			1.338" inside dia		
	Velocity, ft /sec	Velocity head, ft	Head loss, ft /100 ft	Velocity, ft /sec	Velocity head, ft	Head loss, ft /100 ft	Velocity, ft /sec	Velocity head, ft	Head loss, ft /100 ft
4	.63	.006	.166	.73	.01	.233	.913	.013	.404
5	.79	.010	.246	.91	.01	.346	1.14	.020	.601
6	.95	.014	.340	1.09	.02	.478	1.37	.029	.832
7	1.10	.019	.447	1.27	.03	.630	1.60	.040	1.10
8	1.26	.025	.567	1.45	.03	.800	1.83	.052	1.35
9	1.42	.031	.701	1.63	.04	.990	2.05	.065	1.67
10	1.58	.039	.848	1.82	.05	1.20	2.28	.081	2.03
12	1.89	.056	1.18	2.18	.07	1.61	2.74	.116	2.84
14	2.21	.076	1.51	2.54	.10	2.14	3.20	.158	3.78
16	2.52	.099	1.93	2.90	.13	2.74	3.65	.207	4.85
18	2.84	.125	2.40	3.27	.17	3.41	4.11	.262	6.04
20	3.15	.154	2.92	3.63	.20	4.15	4.56	.323	7.36
22	3.47	.187	3.48	3.99	.25	4.96	5.02	.391	8.81
24	3.78	.222	4.10	4.36	.30	5.84	5.48	.465	10.4
26	4.10	.261	4.76	4.72	.35	6.80	5.93	.546	12.1
28	4.41	.303	5.47	5.08	.40	7.82	6.39	.634	13.9
30	4.73	.347	6.23	5.45	.46	8.91	6.85	.727	15.9
32	5.04	.395	7.04	5.81	.52	10.1	7.30	.828	18.0
34	5.36	.446	7.90	6.17	.59	11.3	7.76	.934	20.2
36	5.67	.500	8.80	6.54	.66	12.6	8.22	1.05	22.5
38	5.99	.577	9.76	6.90	.74	14.0	8.67	1.17	25.0
40	6.30	.618	10.8	7.26	.82	15.4	9.13	1.29	27.6
42	6.62	.681	11.8	7.63	.90	16.9	9.58	1.43	30.3
44	6.93	.747	12.9	7.99	.99	18.5	10.04	1.57	33.1
46	7.25	.817	14.0	8.35	1.08	20.1	10.50	1.71	36.1
48	7.56	.889	15.2	8.72	1.18	21.8	10.95	1.86	39.2
50	7.88	.965	16.5	9.08	1.28	23.6	11.41	2.02	42.4
55	8.67	1.17	19.8	9.99	1.55	28.4	12.55	2.45	51.0
60	9.46	1.39	23.4	10.9	1.8	33.6	13.69	2.91	60.4
65	10.24	1.63	27.3	11.8	2.2	39.2	14.83	3.41	70.6
70	11.03	1.89	31.5	12.7	2.5	45.3	15.97	3.96	81.5
75	11.8	2.17	36.0	13.6	2.9	51.8	17.11	4.55	93.2
80	12.6	2.47	40.8	14.5	3.3	58.7	18.25	5.17	106
85	13.4	2.79	45.9	15.4	3.7	66.0	19.40	5.84	119
90	14.2	3.13	51.3	16.3	4.1	73.8	20.54	6.55	133
95	15.0	3.48	57.0	17.2	4.6	82.0	21.68	7.29	148
100	15.8	3.86	63.0	18.2	5.1	90.7	22.82	8.08	164
110	17.3	4.67	75.8	20.0	6.2	109.3	25.10	9.78	197
120	18.9	5.56	89.9	21.8	7.4	129.6	27.38	11.6	234
130	20.5	6.52	105	23.6	8.7	151.6	29.66	13.7	274
140	22.1	7.56	122	25.4	10.0	175			
150	23.6	8.68	139	27.2	11.5	201			
160	25.2	9.88	158	29.0	13.1	228			
170	26.8	11.15	178	30.9	14.8	257			
180	28.4	12.50	199	32.7	16.6	288			

Note: No allowance has been made for age, difference in diameter, or any abnormal condition of interior surface. Any factor of safety must be estimated from the local conditions and the requirements of each particular installation. It is recommended that for most commercial design purposes a safety factor of 15 to 20% be added to the values in the tables.

Reprinted with permission from Cameron Hydraulic *Data Book*.

Friction of Water – New Steel Pipe
(Based on Darcy's Formula)
2 Inch

Flow, U.S. gal per min	Standard wt steel – sch 40			Extra strong steel – sch 80			Steel – Schedule 160		
	2.067" inside dia			1.939" inside dia			1.687" inside dia		
	Velocity, ft /sec	Velocity head, ft	Head loss, ft /100 ft	Velocity, ft /sec	Velocity head, ft	Head loss, ft /100 ft	Velocity, ft /sec	Velocity head, ft	Head loss, ft /100 ft
5	.478	.004	.074	.54	.00	.101	.718	.008	.197
6	.574	.005	.102	.65	.01	.139	.861	.012	.271
7	.669	.007	.134	.76	.01	.182	1.01	.016	.357
8	.765	.009	.170	.87	.01	.231	1.15	.020	.452
9	.860	.012	.209	.98	.01	.285	1.29	.026	.559
10	.956	.014	.252	1.09	.02	.343	1.44	.032	.675
12	1.15	.021	.349	1.30	.03	.476	1.72	.046	.938
14	1.34	.028	.461	1.52	.04	.629	2.01	.063	1.20
16	1.53	.036	.586	1.74	.05	.800	2.30	.082	1.53
18	1.72	.046	.725	1.96	.06	.991	2.58	.104	1.90
20	1.91	.057	.878	2.17	.07	1.16	2.87	.128	2.31
22	2.10	.069	1.05	2.39	.09	1.38	3.16	.155	2.76
24	2.29	.082	1.18	2.61	.11	1.62	3.45	.184	3.25
26	2.49	.096	1.37	2.83	.12	1.88	3.73	.216	3.77
28	2.68	.111	1.57	3.04	.14	2.16	4.02	.251	4.33
30	2.87	.128	1.82	3.26	.17	2.46	4.31	.288	4.93
35	3.35	.174	2.38	3.80	.22	3.28	5.02	.392	6.59
40	3.82	.227	3.06	4.35	.29	4.21	5.74	.512	8.49
45	4.30	.288	3.82	4.89	.37	5.26	6.46	.648	10.6
50	4.78	.355	4.66	5.43	.46	6.42	7.18	.799	13.0
55	5.26	.430	5.58	5.98	.56	7.70	7.89	.967	15.6
60	5.74	.511	6.58	6.52	.66	9.09	8.61	1.15	18.4
65	6.21	.600	7.66	7.06	.77	10.59	9.33	1.35	21.5
70	6.69	.696	8.82	7.61	.90	12.2	10.05	1.57	24.8
75	7.17	.799	10.1	8.15	1.03	13.9	10.77	1.80	28.3
80	7.65	.909	11.4	8.69	1.17	15.8	11.48	2.05	32.1
85	8.13	1.03	12.8	9.03	1.27	17.7	12.20	2.31	36.1
90	8.60	1.15	14.3	9.78	1.49	19.8	12.92	2.59	40.3
95	9.08	1.28	15.9	10.3	1.6	22.0	13.64	2.89	44.8
100	9.56	1.42	17.5	10.9	1.8	24.3	14.35	3.20	49.5
110	10.52	1.72	21.0	12.0	2.2	29.2	15.79	3.87	59.6
120	11.5	2.05	24.9	13.0	2.6	34.5	17.22	4.61	70.6
130	12.4	2.40	29.1	14.1	3.1	40.3	18.66	5.40	82.6
140	13.4	2.78	33.6	15.2	3.6	46.6	20.10	6.27	95.5
150	14.3	3.20	38.4	16.3	4.1	53.3	21.53	7.20	109
160	15.3	3.64	43.5	17.4	4.7	60.5	22.97	8.19	124
170	16.3	4.11	49.0	18.5	5.3	68.1	24.40	9.24	140
180	17.2	4.60	54.8	19.6	6.0	76.1	25.84	10.36	156
190	18.2	5.13	60.9	20.6	6.6	84.6	27.27	11.54	174
200	19.1	5.68	67.3	21.7	7.3	93.6	28.71	12.79	192
220	21.0	6.88	81.1	23.9	8.9	113			
240	22.9	8.18	96.2	26.9	10.6	134			
260	24.9	9.60	113	28.3	12.4	157			
280	26.8	11.14	130	30.4	14.4	181			
300	28.7	12.8	149	32.6	16.5	208			

Note: No allowance has been made for age, difference in diameter, or any abnormal condition of interior surface. Any factor of safety must be estimated from the local conditions and the requirements of each particular installation. It is recommended that for most commercial design purposes a safety factor of 15 to 20% be added to the values in the tables.

Reprinted with permission from Cameron Hydraulic *Data Book*.

Friction of Water – New Steel Pipe
(Based on Darcy's Formula)
2½ Inch

Flow, U.S. gal per min	Standard wt steel – sch 40			Extra strong steel – sch 80			Steel – Schedule 160		
	2.469" inside dia			2.323" inside dia			2.125" inside dia		
	Velocity, ft /sec	Velocity head, ft	Head loss, ft /100 ft	Velocity, ft /sec	Velocity head, ft	Head loss, ft /100 ft	Velocity, ft /sec	Velocity head, ft	Head loss, ft /100 ft
8	.536	.005	.072	.61	.01	.097	.724	.008	.149
10	.670	.007	.107	.76	.01	.144	.905	.013	.221
12	.804	.010	.148	.91	.01	.199	1.09	.018	.305
14	.938	.014	.195	1.06	.02	.261	1.27	.025	.403
16	1.07	.018	.247	1.21	.02	.332	1.45	.033	.512
18	1.21	.023	.305	1.36	.03	.411	1.63	.041	.634
20	1.34	.028	.369	1.51	.04	.497	1.81	.051	.767
22	1.47	.034	.438	1.67	.04	.590	1.99	.061	.912
24	1.61	.040	.513	1.82	.05	.691	2.17	.073	1.03
26	1.74	.047	.593	1.97	.06	.800	2.35	.086	1.20
28	1.88	.055	.679	2.12	.07	.915	2.53	.100	1.37
30	2.01	.063	.770	2.27	.08	1.00	2.71	.114	1.56
35	2.35	.086	0.99	2.65	.11	1.33	3.17	.156	2.08
40	2.68	.112	1.26	3.03	.14	1.71	3.62	.203	2.66
45	3.02	.141	1.57	3.41	.18	2.13	4.07	.257	3.32
50	3.35	.174	1.91	3.79	.22	2.59	4.52	.318	4.05
55	3.69	.211	2.28	4.16	.27	3.10	4.98	.384	4.85
60	4.02	.251	2.69	4.54	.32	3.65	5.43	.457	5.72
65	4.36	.295	3.13	4.92	.38	4.25	5.88	.537	6.66
70	4.69	.342	3.60	5.30	.44	4.89	6.33	.622	7.67
75	5.03	.393	4.10	5.68	.50	5.58	6.79	.714	8.75
80	5.36	.447	4.64	6.05	.57	6.31	7.24	.813	9.90
85	5.70	.504	5.20	6.43	.64	7.08	7.69	.918	11.1
90	6.03	.565	5.80	6.81	.72	7.89	8.14	1.03	12.4
95	6.37	.630	6.43	7.19	.80	8.76	8.59	1.15	13.8
100	6.70	.698	7.09	7.57	.89	9.66	9.05	1.27	15.2
110	7.37	.844	8.51	8.33	1.08	11.6	9.95	1.54	18.3
120	8.04	1.00	10.1	9.08	1.28	13.7	10.86	1.83	21.6
130	8.71	1.18	11.7	9.84	1.50	16.0	11.76	2.15	25.2
140	9.38	1.37	13.5	10.6	1.7	18.5	12.67	2.49	29.1
150	10.05	1.57	15.5	11.3	2.0	21.1	13.57	2.86	33.3
160	10.7	1.79	17.5	12.1	2.3	23.9	14.47	3.25	37.8
170	11.4	2.02	19.7	12.9	2.6	26.9	15.38	3.67	42.5
180	12.1	2.26	22.0	13.6	2.9	30.1	16.28	4.12	47.5
190	12.7	2.52	24.4	14.4	3.2	33.4	17.19	4.59	52.8
200	13.4	2.79	27.0	15.1	3.5	36.9	18.09	5.08	58.4
220	14.7	3.38	32.5	16.7	4.3	44.4	19.90	6.15	70.3
240	16.1	4.02	38.5	18.2	5.1	52.7	21.71	7.32	83.4
260	17.4	4.72	45.0	19.7	6.0	61.6	23.52	8.59	97.6
280	18.8	5.47	52.3	21.2	7.0	71.2	25.33	9.96	113
300	20.1	6.28	59.6	22.7	8.0	81.6	27.14	11.43	129
350	23.5	8.55	80.6	26.5	10.9	110	31.66	15.56	175
400	26.8	11.2	105	30.3	14.3	144	36.19	20.32	228
450	30.2	14.1	132	34.1	18.1	181	40.71	25.72	288
500	33.5	17.4	163	37.9	22.3	223	45.23	31.75	354

Note: No allowance has been made for age, difference in diameter, or any abnormal condition of interior surface. Any factor of safety must be estimated from the local conditions and the requirements of each particular installation. It is recommended that for most commercial design purposes a safety factor of 15 to 20% be added to the values in the tables.

Reprinted with permission from Cameron Hydraulic Data Book.

Friction of Water
Asphalt-Dipped Cast-Iron and New Steel Pipe
(Based on Darcy's Formula)
3 Inch

Flow	Asphalt-dipped cast iron			Std wt steel sch 40			Extra strong steel sch 80			Steel— schedule 160		
	3.0" inside dia			3.068" inside dia			2.900" inside dia			2.624" inside dia		
U.S. gal per min	Velocity, ft per sec	Velocity head, ft	Head loss, ft per 100 ft	Velocity, ft per sec	Velocity head, ft	Head loss, ft per 100 ft	Velocity, ft per sec	Velocity head, ft	Head loss, ft per 100 ft	Velocity, ft per sec	Velocity head, ft	Head loss, ft per 100 ft
10	.454	.00	.042	.434	.003	.038	.49	.00	.050	.593	.005	.080
15	.681	.01	.088	.651	.007	.077	.73	.01	.101	.890	.012	.164
20	.908	.01	.149	.868	.012	.129	.97	.02	.169	1.19	.022	.275
25	1.13	.02	.225	1.09	.018	.192	1.21	.02	.253	1.48	.034	.411
30	1.36	.03	.316	1.30	.026	.267	1.45	.03	.351	1.78	.049	.572
35	1.59	.04	.421	1.52	.036	.353	1.70	.04	.464	2.08	.067	.757
40	1.82	.05	.541	1.74	.047	.449	1.94	.06	.592	2.37	.087	.933
45	2.04	.06	.676	1.95	.059	.557	2.18	.07	.734	2.67	.111	1.16
50	2.27	.08	.825	2.17	.073	.676	2.43	.09	.860	2.97	.137	1.41
55	2.50	.10	.990	2.39	.089	.776	2.67	.11	1.03	3.26	.165	1.69
60	2.72	.12	1.17	2.60	.105	.912	2.91	.13	1.21	3.56	.197	1.99
65	2.95	.14	1.36	2.82	.124	1.06	3.16	.15	1.40	3.86	.231	2.31
70	3.18	.16	1.57	3.04	.143	1.22	3.40	.18	1.61	4.15	.268	2.65
75	3.40	.18	1.79	3.25	.165	1.38	3.64	.21	1.83	4.45	.307	3.02
80	3.63	.21	2.03	3.47	.187	1.56	3.88	.23	2.07	4.75	.350	3.41
85	3.86	.23	2.28	3.69	.211	1.75	4.12	.26	2.31	5.04	.395	3.83
90	4.08	.26	2.55	3.91	.237	1.95	4.37	.29	2.58	5.34	.443	4.27
95	4.31	.29	2.83	4.12	.264	2.16	4.61	.33	2.86	5.63	.493	4.73
100	4.54	.32	3.12	4.34	.293	2.37	4.85	.36	3.15	5.93	.546	5.21
110	4.99	.39	3.75	4.77	.354	2.84	5.33	.44	3.77	6.53	.661	6.25
120	5.45	.46	4.45	5.21	.421	3.35	5.81	.52	4.45	7.12	.787	7.38
130	5.90	.54	5.19	5.64	.495	3.90	6.30	.62	5.19	7.71	.923	8.61
140	6.35	.63	6.00	6.08	.574	4.50	6.79	.71	5.98	8.31	1.07	9.92
150	6.81	.72	6.87	6.51	.659	5.13	7.28	.82	6.82	8.90	1.23	11.3
160	7.26	.82	7.79	6.94	.749	5.80	7.76	.93	7.72	9.49	1.40	12.8
180	8.17	1.04	9.81	7.81	.948	7.27	8.72	1.01	9.68	10.68	1.77	16.1
200	9.08	1.28	12.1	8.68	1.17	8.90	9.70	1.46	11.86	11.87	2.19	19.8
220	9.98	1.55	14.5	9.55	1.42	10.7	10.7	1.78	14.26	13.05	2.64	23.8
240	10.9	1.84	17.3	10.4	1.69	12.7	11.6	2.07	16.88	14.24	3.15	28.2
260	11.8	2.16	20.2	11.3	1.98	14.8	12.6	2.46	19.71	15.43	3.69	32.9
280	12.7	2.51	23.4	12.2	2.29	17.1	13.6	2.88	22.77	16.61	4.28	38.0
300	13.6	2.88	26.8	13.0	2.63	19.5	14.5	3.26	26.04	17.80	4.92	43.5
320	14.5	3.28	30.4	13.9	3.00	22.1	15.5	3.77	29.53	18.99	5.59	49.4
340	15.4	3.70	34.3	14.8	3.38	24.9	16.5	4.22	33.24	20.17	6.32	55.6
360	16.3	4.15	38.4	15.6	3.79	27.8	17.5	4.73	37.16	21.36	7.08	62.2
380	17.2	4.62	42.7	16.5	4.23	30.9	18.4	5.27	41.31	22.55	7.89	69.2
400	18.2	5.12	47.3	17.4	4.68	34.2	19.4	5.81	45.67	23.73	8.74	76.5
420	19.1	5.65	52.1	18.2	5.16	37.6	20.4	6.43	50.25	24.92	9.64	84.2
440	20.0	6.20	57.1	19.1	5.67	41.2	21.4	7.13	55.05	26.11	10.58	92.2
460	20.9	6.77	62.4	20.0	6.19	44.9	22.3	7.75	60.06	27.29	11.56	101
480	21.8	7.38	67.9	20.8	6.74	48.8	23.3	8.37	65.30	28.48	12.59	109
500	22.7	8.00	73.6	21.7	7.32	52.9	24.2	9.15	70.75	29.66	13.66	119
550	25.0	9.68	88.9	23.9	8.85	63.8	26.7	11.1	85.33	32.63	16.53	143
600	27.2	11.5	106	26.0	10.5	75.7	29.1	13.1	101	35.60	19.67	170
650	29.5	13.5	124	28.2	12.4	88.6	31.6	15.5	119	38.56	23.08	199

Note: No allowance has been made for age, difference in diameter, or any abnormal condition of interior surface. Any factor of safety must be estimated from the local conditions and the requirements of each particular installation. It is recommended that for most commercial design purposes a safety factor of 15 to 20% be added to the values in the tables.

Reprinted with permission from Cameron Hydraulic *Data Book.*

Part Two: Friction Losses in Pipe Fittings

Friction of Water
Friction Loss in Pipe Fittings

Resistance coefficient K $\left(\text{use in formula } h_f = \dfrac{K\,V^2}{2g}\right)$

Note: Fittings are standard with full openings.

Fitting	L/D	Nominal pipe size											
		½	¾	1	1¼	1½	2	2½–3	4	6	8–10	12–16	18–24
		K value											
Gate Valves	8	0.22	0.20	0.18	0.18	0.15	0.15	0.14	0.14	0.12	0.11	0.10	0.10
Globe Valves	340	9.2	8.5	7.8	7.5	7.1	6.5	6.1	5.8	5.1	4.8	4.4	4.1
Angle Valves	55	1.48	1.38	1.27	1.21	1.16	1.05	0.99	0.94	0.83	0.77	0.72	0.66
Angle Valves	150	4.05	3.75	3.45	3.30	3.15	2.85	2.70	2.55	2.25	2.10	1.95	1.80
Ball Valves	3	0.08	0.08	0.07	0.07	0.06	0.06	0.05	0.05	0.05	0.04	0.04	0.04

Calculated from data in Crane Co. Technical Paper No. 410.
Reprinted with permission from Cameron Hydraulic *Data Book*.

Note: Fittings are standard with full openings.

Friction of Water
Friction Loss in Pipe Fittings

Resistance coefficient K $\left(\text{use in formula } h_f = \dfrac{K V^2}{2g}\right)$

Fitting	L/D	Nominal pipe size											
		1/2	3/4	1	1¼	1½	2	2½–3	4	6	8–10	12–16	18–24
		K value											
Butterfly Valve							0.86	0.81	0.77	0.68	0.63	0.35	0.30
Plug Valve straightway	18	0.49	0.45	0.41	0.40	0.38	0.34	0.32	0.31	0.27	0.25	0.23	0.22
Plug Valv 3-way thru-flo	30	0.81	0.75	0.69	0.66	0.63	0.57	0.54	0.51	0.45	0.42	0.39	0.36
Plug Valve branch-flo	90	2.43	2.25	2.07	1.98	1.89	1.71	1.62	1.53	1.35	1.26	1.17	1.08
Standard elbow 90°	30	0.81	0.75	0.69	0.66	0.63	0.57	0.54	0.51	0.45	0.42	0.39	0.36
45°	16	0.43	0.40	0.37	0.35	0.34	0.30	0.29	0.27	0.24	0.22	0.21	0.19
long radius 90°	16	0.43	0.40	0.37	0.35	0.34	0.30	0.29	0.27	0.24	0.22	0.21	0.19

Calculated from data in Crane Co. Technical Paper No. 410.
Reprinted with permission from Cameron Hydraulic *Data Book*.

Friction of Water
Friction Loss in Pipe Fittings

Resistance coefficient K $\left(\text{use in formula } h_f = \dfrac{KV^2}{2g}\right)$

Note: Fittings are standard with full openings.

Fitting	Type of bend	L/D	½	¾	1	1¼	1½	2	2½–3	4	6	8–10	12–16	18–24
			K value											
Close Return Bend		50	1.35	1.25	1.15	1.10	1.05	0.95	0.90	0.85	0.75	0.70	0.65	0.60
Standard Tee	thru flo	20	0.54	0.50	0.46	0.44	0.42	0.38	0.36	0.34	0.30	0.28	0.26	0.24
	thru branch	60	1.62	1.50	1.38	1.32	1.26	1.14	1.08	1.02	0.90	0.84	0.78	0.72
90° Bends. Pipe bends, flanged elbows, butt-welded elbows	r/d = 1	20	0.54	0.50	0.46	0.44	0.42	0.38	0.36	0.34	0.30	0.28	0.26	0.24
	r/d = 2	12	0.32	0.30	0.28	0.26	0.25	0.23	0.22	0.20	0.18	0.17	0.16	0.14
	r/d = 3	12	0.32	0.30	0.28	0.26	0.25	0.23	0.22	0.20	0.18	0.17	0.16	0.14
	r/d = 4	14	0.38	0.35	0.32	0.31	0.29	0.27	0.25	0.24	0.21	0.20	0.18	0.17
	r/d = 6	17	0.46	0.43	0.39	0.37	0.36	0.32	0.31	0.29	0.26	0.24	0.22	0.20
	r/d = 8	24	0.65	0.60	0.55	0.53	0.50	0.46	0.43	0.41	0.36	0.34	0.31	0.29
	r/d = 10	30	0.81	0.75	0.69	0.66	0.63	0.57	0.54	0.51	0.45	0.42	0.39	0.36
	r/d = 12	34	0.92	0.85	0.78	0.75	0.71	0.65	0.61	0.58	0.51	0.48	0.44	0.41
	r/d = 14	38	1.03	0.95	0.87	0.84	0.80	0.72	0.68	0.65	0.57	0.53	0.49	0.46
	r/d = 16	42	1.13	1.05	0.97	0.92	0.88	0.80	0.76	0.71	0.63	0.59	0.55	0.50
	r/d = 18	46	1.24	1.15	1.06	1.01	0.97	0.87	0.83	0.78	0.69	0.64	0.60	0.55
	r/d = 20	50	1.35	1.25	1.15	1.10	1.05	0.95	0.90	0.85	0.75	0.70	0.65	0.60
Mitre Bends	α = 0°	2	0.05	0.05	0.05	0.04	0.04	0.04	0.04	0.03	0.03	0.03	0.03	0.02
	α = 15°	4	0.11	0.10	0.09	0.09	0.08	0.08	0.07	0.07	0.06	0.06	0.05	
	α = 30°	8	0.22	0.20	0.18	0.18	0.17	0.15	0.14	0.14	0.12	0.11	0.10	0.10
	α = 45°	15	0.41	0.38	0.35	0.33	0.32	0.29	0.27	0.26	0.23	0.21	0.20	0.18
	α = 60°	25	0.68	0.63	0.58	0.55	0.53	0.48	0.45	0.43	0.38	0.35	0.33	0.30
	α = 75°	40	1.09	1.00	0.92	0.88	0.84	0.76	0.72	0.68	0.60	0.56	0.52	0.48
	α = 90°	60	1.62	1.50	1.38	1.32	1.26	1.14	1.08	1.02	0.90	0.84	0.78	0.72

Nominal pipe size

Calculated from data in Crane Co. Technical Paper No. 410.
Reprinted with permission from Cameron Hydraulic *Data Book*.

Friction of Water
Friction Loss in Pipe Fittings

Resistance coefficient K $\left(\text{use in formula } h_f = \dfrac{K V^2}{2g}\right)$

Note: Fittings are standard with full openings.

Fitting	L/D	Min. velocity for full disc lift — general ft/sec*	Min. velocity for full disc lift — water, ft/sec	Nominal pipe size — K value** ½	¾	1	1¼	1½	2	2½–3	4	6	8–10	12–16	18–24
Stop-check valves	400	55 √V	6.96	10.8	10	9.2	8.8	8.4	7.5	7.2	6.8	6.0	5.6	5.2	4.8
	200	75 √V	9.49	5.4	5	4.6	4.4	4.2	3.8	3.6	3.4	3.0	2.8	2.6	2.4
	350	60 √V	7.59	9.5	8.8	8.1	7.7	7.4	6.7	6.3	6.0	5.3	4.9	4.6	4.2
	300	60 √V	7.59	8.1	7.5	6.9	6.6	6.3	5.7	5.4	5.1	4.5	4.2	3.9	3.6
	55	140 √V	17.7	1.5	1.4	1.3	1.2	1.2	1.1	1.0	.94	.83	.77	.72	.66

Calculated from data in Crane Co. Technical Paper No. 410.   Reprinted with permission from Cameron Hydraulic *Data Book.*

* In these formulas, V is specific volume (ft³/lb).

** These K values are for flow giving full disc lift. K values are higher for low flows giving partial disc lift.

Friction of Water
Friction Loss in Pipe Fittings

Resistance coefficient K $\left(\text{use in formula } h_f = \dfrac{K V^2}{2g}\right)$

Note: Fittings are standard with full openings.

Fitting	L/D	Min. velocity for full disc lift general ft/sec*	water ft/sec	½	¾	1	1¼	1½	2	2½–3	4	6	8–10	12–16	18–24
									K value**						
Swing check valves	100	$35\sqrt{V}$	4.43	2.7	2.5	2.3	2.2	2.1	1.9	1.8	1.7	1.5	1.4	1.3	1.2
	50	$48\sqrt{V}$	6.08	1.4	1.3	1.2	1.1	1.1	1.0	0.9	0.9	.75	.70	.65	.6
Lift check valves	600	$40\sqrt{V}$	5.06	16.2	15	13.8	13.2	12.6	11.4	10.8	10.2	9.0	8.4	7.8	7.2
	55	$140\sqrt{V}$	17.7	1.5	1.4	1.3	1.2	1.2	1.1	1.0	.94	.83	.77	.72	.66
Tilting disc check valve	5°	$80\sqrt{V}$	10.13						.76	.72	.68	.60	.56	.39	.24
	15°	$30\sqrt{V}$	3.80						2.3	2.2	2.0	1.8	1.7	1.2	.72
Foot valve with strainer poppet disc	420	$15\sqrt{V}$	1.90	11.3	10.5	9.7	9.3	8.8	8.0	7.6	7.1	6.3	5.9	5.5	5.0
Foot valve with strainer hinged disc	75	$35\sqrt{V}$	4.43	2.0	1.9	1.7	1.7	1.7	1.4	1.4	1.3	1.1	1.1	1.0	.90

Nominal pipe size

Calculated from data in Crane Co. Technical Paper No. 410. Reprinted with permission from Cameron Hydraulic *Data Book*.

* In these formulas, V is specific volume (ft³/lb).

** These K values for flow giving full disc lift. K values are higher for low flows giving partial disc lift.

Friction of Water
Friction Loss in Pipe Fittings

Resistance coefficient K $\left(\text{use in formula } h_f = \dfrac{K V^2}{2g}\right)$

Fitting	Description	All pipe sizes
		K value
Pipe exit	projecting sharp edged rounded	1.0
Pipe entrance	inward projecting	0.78
Pipe entrance flush	sharp edged	0.5
	r/d = 0.02	0.28
	r/d = 0.04	0.24
	r/d = 0.06	0.15
	r/d = 0.10	0.09
	r/d = 0.15 & up	0.04

From Crane Co. Technical Paper No. 410.

Reprinted with permission from Cameron Hydraulic *Data Book*.

Part Three: Corresponding Pressure Table

Corresponding pressure in pounds per square inch (psi) to corresponding head in feet

psi	0	1	2	3	4	5	6	7	8	9
0		2.3	4.6	6.9	9.2	11.6	13.9	16.2	18.5	20.8
10	23.1	25.4	27.7	30.0	32.3	34.7	37.0	39.3	41.6	43.9
20	46.2	48.5	50.8	53.1	55.4	57.8	60.1	62.4	64.7	67.0
30	69.3	71.6	73.9	76.2	78.5	80.9	83.2	85.5	87.8	90.1
40	92.4	94.7	97.0	99.3	101.6	104.0	106.3	108.6	110.9	113.2
50	115.5	117.8	120.1	122.4	124.7	127.1	129.4	131.7	134.0	136.3
60	138.6	140.9	143.2	145.5	147.8	150.2	152.5	154.8	157.1	159.4
70	161.7	164.0	166.3	168.6	170.9	173.3	175.6	177.9	180.2	182.5
80	184.8	187.1	189.4	191.7	194.0	196.4	198.7	201.0	203.3	205.6
90	207.9	210.2	212.5	214.8	217.7	219.5	221.8	224.1	226.4	228.7
100	231.0	233.3	235.6	237.9	240.2	242.6	244.9	247.2	249.5	251.8
110	254.1	256.4	258.7	261.0	263.3	265.7	268.0	270.3	272.6	274.9
120	277.2	279.5	281.8	284.1	286.4	288.8	291.1	293.4	295.7	298.0
130	300.3	302.6	304.9	307.2	309.5	311.9	314.2	316.5	318.8	321.1
140	323.4	325.7	328.0	330.3	332.6	335.0	337.3	339.6	341.9	344.2
150	346.5	348.8	351.1	353.4	355.7	358.1	360.4	362.7	365.0	367.3
160	369.6	371.9	374.2	376.5	378.8	381.2	383.5	385.8	388.1	390.4
170	392.7	395.0	397.3	399.6	401.9	404.3	406.6	408.9	411.2	413.5
180	415.8	418.1	420.4	422.7	425.0	427.4	429.7	432.0	434.3	436.6
190	438.9	441.2	443.5	445.8	448.1	450.5	452.8	455.1	457.4	459.7
200	462.0	464.3	466.6	468.9	471.2	473.6	475.9	478.2	480.5	482.8
210	485.1	487.4	489.7	492.0	494.3	496.7	499.0	501.3	503.6	505.9
220	508.2	510.5	512.8	515.1	517.4	519.8	522.1	524.4	526.7	529.0
230	531.3	533.6	535.9	538.2	540.5	542.9	545.2	547.5	549.8	552.1
240	554.4	556.7	559.0	561.3	563.6	566.0	568.3	570.6	572.9	575.2
250	577.5	579.8	582.1	584.4	586.7	589.1	591.4	593.7	596.0	598.3

To read the table, use the following example as a guide.

For 13 psi, go horizontal from 10 at the left and read under 3 vertically.
Answer is 30 ft (corresponding to 13 psi).

Part Four: Symbols

Name	Symbol
NEW PIPING	· ———————————
EXISTING PIPING (ABOVE GRADE)	—— **EXIST** **SP** ——
UNDERGROUND PIPING (NEW)	++++++++++++++++++++++
UNDERGROUND PIPING (EXISTING)	⤬⤬⤬⤬⤬⤬⤬⤬⤬⤬⤬⤬⤬
SPRINKLER LINE	——— **SP** ———
LINE TO BE REMOVED	⤬ ⤬ ⤬
FIRE STANDPIPE	——— **F** ———
FLUSH MOUNTED	——⊗——
UPRIGHT SPRINKLER HEAD	——○——
PENDENT SPRINKLER HEAD	——⨂——
DROP OR RISE	——○——
RISE OR DROP	——(——
SPRINKLER HEAD WITH PROTECTIVE CAGE (NORMALLY PENDENT)	—○c—

Name	Symbol
INCREASER OR REDUCER	
1" INSPECTOR'S TEST CONNECTION	
CAPPED LINE	
FLANGED LINE	
ELBOW (45°)	
HIGH TEMPERATURE HEAD (NEAR HEATER)	(°)
UNION	
BELL OR GONG	F
PUMP TEST HEADER	
SIDEWALL SPRINKLER HEAD	
HORN	F
MATCH LINE	

Name	Symbol
GATE VALVE (GV)	
GLOBE VALVE (GL.V)	
CHECK VALVE (CK.V)	
BACKFLOW PREVENTER (BFP)	BFP
BUTTERFLY VALVE (BV)	
THREE-WAY VALVE	
SOLENOID VALVE	S
MOTOR OPERATED VALVE	M
FIRE HOSE VALVE (FHV)	
SPRINKLER ALARM VALVE WITH TRIMMINGS (AV)	
DRY PIPE VALVE (DPV)	D
PRE-ACTION VALVE (PAV)	P

Name	Symbol
CONTROL VALVE IN VALVE BOX	
POST INDICATOR VALVE	
CONTROL VALVE UNDERGROUND	
AIR VENT (AUTOMATIC)	
CHAIN OPERATED VALVE	**COV**
DELUGE VALVE	
HEAT SENSING DEVICE	
FIRE DEPARTMENT INLET, TWO WAY WITH DOUBLE CLAPPER	
FIRE HYDRANT	
SIAMESE (FREE STANDING TYPE)	
WATER FLOW SWITCH (WFS)	
FIRE PUMP	

Name	Symbol
SIAMESE (WALL TYPE)	
GAUGE	
TAMPER SWITCH	TS
FIRE HOSE RACK	
FIRE EXTINGUISHER	E
METER (FIRE/WATER)	M
MANUAL FIRE ALARM PULL STATION	F
MUNICIPAL FIRE ALARM STATION	F
FIRE DETECTOR (DUCT)	F$_D$
FIRE DETECTOR (IONIZATION)	F$_I$
FIRE DETECTOR (INFRARED)	F$_{IR}$
FIRE DETECTOR (PHOTOELECTRIC)	F$_P$

Name	Symbol
FIRE DETECTOR (THERMAL)	F_T
FIRE DETECTOR (ULTRAVIOLET)	F_UV
PRESSURE SWITCH	PS
FLOW SWITCH	FS
PRESSURE INDICATOR	PI
TEMPERATURE ELEMENT	TE
TEMPERATURE CONTROL VALVE	TCV
"BREAK GLASS" STATION	MS
BUILDING FIRE ALARM CONTROL PANEL	FACP-B
CARBON DIOXIDE (CO_2) CONTROL PANEL	CCP
PRE-ACTION SPRINKLER SYSTEM CONTROL PANEL	SCP
FIRE HOSE CABINET	FHC

Appendix C

Part One: Pre-Installation Designer's and Contractor's Information List

A. Total Flooding CO_2 System

1. Name of hazards to be protected and dimensions (length, width, and height) necessary to calculate volumes. Create accurate sketches.
2. Hazard location, i.e., building number, floor, area, department.
3. Identify key contact person, title, phone number, best times to be reached.
4. Describe machinery, equipment, and operation in each space in detail.
5. Determine ambient and process temperatures, rate of rise.
6. Identify flammable materials; secure Material Safety Data Sheets for determination of agent concentration levels.
7. Show number, size, location, and a full description of all openings, such as doors, windows, skylights, ventilation, and conveyor openings. Identify as:
   A. Normally closed;
   B. To be closed by system actuation;
   C. Cannot be closed. Who is responsible for automatic closing devices and dampers?
8. Ventilation—Forced or natural? If forced ventilation is used, determine details of ductwork, damper locations intake, exhaust locations, electrical data, and wind-down

time for fans. Is air proprietary to that space? Who will be responsible for electrical work?

9. Obtain horsepower, voltage, and current of motors and other equipment to be shut down. Is there a holding coil circuit?

10. If fuel is gas or oil, determine size of feed line. Provide fusible link valve.

11. If vapors are drawn into a manifold system involving other processes, determine effect on these other operations.

12. Occupancy—Are people normally working in the protected area? Is discharge delay required? If area is normally unoccupied, what notification signal will prevent someone from inadvertently entering flooded space?

13. Who is responsible for training, marking of egress routes, supplementary breathing apparatus?

14. Is explosion-proof equipment required in hazard area?

15. Is alarm tie-in to local fire department required by the authority?

16. Cylinder location—Determine placement. This will affect amount of pipe and detection lines needed. Also determine location of remote manual pull stations.

17. Describe ceiling construction. Type of joists, beams, and sloped or peaked ceiling will affect nozzle, piping, and detector spacing, piping layout, and attachment.

18. Indicate location and size of machinery or equipment in the space that might interfere with nozzle discharge pattern.

19. When two or more hazards are being protected, show location of all hazards in relation to each other.

20. Note construction surrounding the hazard: wood, steel, concrete.

21. Indicate where discharge pipe, detection lines, and remote-control cable can be run.

22. Can work be done in unrestricted fashion during normal business hours or must night/weekend hours be used for timely completion?

23. Do the normal operations pose any danger to those installing the system, for example, explosion dangers posed by the use of electrical tools, risks from work around running machines, scaffold or man-lift work?

24. Is union labor required?

B. CO$_2$ Systems Local Application

1. Name of hazards to be protected and dimensions necessary to determine coverage either by volume or area.

2. Hazard locations, i.e., building number, floor, area, department.

3. Identify key contact person, title, phone number, best time to be reached.

4. Describe machinery, equipment, and operation in detail.

5. Determine ambient and process temperatures, rate of rise.

6. Identify flammable materials; secure Material Safety Data Sheets. Is material or atmosphere corrosive?

7. Does hazard extend to other areas, such as over to other floors, conveyors, drip pans, through unclosable openings?

8. Include in description (drawings) any openings by size, location, and whether they are:

    A. Normally open or closed during operation;

    B. To be closed by system actuation;

    C. Cannot be closed. Who is responsible for automatic closing devices and dampers?

9. Nozzle locations—Determine where nozzles can be placed without interfering with normal operations, yet where no obstructions block the flow of carbon dioxide to the hazards, and where no one is tempted to move the nozzles to conduct normal operations—for example, to remove and replace rollers, perform routine machine maintenance, load stock.

10. Ventilation—Draft natural or forced? Does the hazard incorporate a solvent recovery unit? Determine details of ductwork, damper locations, electrical data, wind-down times for fans, and relationship to manifold system. De-

termine effect on other operations. Who will be responsible for electrical work?

11. If fuel is gas or oil, determine size of feed line. Provide fusible link valve.
12. How will safety of personnel be maintained?
13. Is explosion-proof equipment required in hazard area?
14. Examine current, voltage, and horsepower of motors, heaters, and other electrical equipment to be shut down. Is there a holding coil circuit?
15. Who is responsible for training of personnel?
16. Is alarm tie-in to local fire department required by the authority having jurisdiction?
17. Cylinder location—Determine placement. This will affect the amount of pipe and detection lines needed. Also determine the location of remote manual pull stations.
18. How is distribution piping to be attached to machines? What is the effect of vibration on detection and agent lines?
19. Are there any deep-seated hazards that require an extended discharge of agent?
20. Is there the potential for a flash fire from adjacent operations, such as ovens?
21. Determine the minimum freeboard space between top of liquid level and top of holding tank. Show size, shape, and description of dipped material.
22. Examine need for coverage of drainboard area and run-off area. Can fire spread to adjoining areas? Can tankside nozzles be used? Will placement and removal of work strike nozzles?
23. Observe the process of machine operation to determine whether any changes are imposed, employee habits that affect fire hazard exist, or diminution of protection may be caused in the normal course of use.
24. Can installation of fire-protection system be done in unrestricted manner during normal business hours or must night/weekend hours be used for timely completion?
25. Are any dangers posed to personnel installing the fire-protection system, such as explosion dangers from use

of electric tools, risk from work around running machines, or dangers presented by unforeseen events?

26. Is union labor needed?

Part Two: Sample of Fire-Protection Notes

Small Project—General Notes (to be included on drawings when a detailed specification is not prepared)

1. Sprinkler system shall be installed in strict conformity with all latest requirements of the state building code, the local fire department, and the National Fire Protection Association (NFPA).

2. Sprinkler control valves shall be 175 psi, UL approved, OS&Y gate valves.

3. Control valve locations shall be as indicated by approved arrows and signs.

4. Control valves more than 7'-O" above the floor shall be accessible by means of permanent iron ladders as per NFPA No. 13.

5. Sprinkler pipe system shall be marked and identified in accordance with NFPA No. 13.

6. Control valves shall be sealed in open position as per NFPA No. 13.

7. Alarm-actuating devices shall be Fire Department and UL approved type as per NFPA No. 13.

8. Alarm-actuating devices shall be connected to supervised alarm system in accordance with NFPA No. 13.

9. Inspector's test connection shall be in accordance with NFPA No. 13.

10. Pipe shall be protected against freezing in accordance with NFPA No. 13.

11. Pipe hangers shall be in accordance with NFPA No. 13.

12. Test of system shall be in accordance with NFPA No. 13.

13. Spare sprinklers shall be stocked in accordance with NFPA No. 13.

14. Water lines shall be installed so as to not interfere with the operation of equipment or other appurtenances.

15. Contractor shall perform all excavation with caution so as to not to damage or disrupt service of existing sewers, gas mains, electrical cables, or other subsurface utilities or appurtenances.

Alternative Sample of Fire-Protection Notes

1. Sprinklers shall be ½ in. orifice pendent type where ceiling is installed, and upright type where no ceiling is installed.
2. All sprinkler system components shall be of listed and approved type.
3. Fittings shall be as defined in NFPA No. 13.
4. All hanger components shall be of listed and approved type.
5. Piping shall be hung from structural elements. Piping shall not be hung from ductwork or from bottom chord of open web steel bar joists.
6. Piping shall be black steel, Schedule 40, ASTM 53.
7. Pipe sizes shall be determined based on pipe schedule NFPA No. 13 (or based on hydraulic calculation). Specifier: select one.
8. Before installation, drawing shall be approved by the owner's fire-insurance company and the authorities having jurisdiction.
9. Contractor shall run and pay for all tests required to ensure that the system has adequate flow and pressure.
10. Where practical, piping shall be located within partitions and supported by the steel studs.
11. Contractor shall furnish as-built drawings.
12. Contractor shall be responsible during installation and testing for any damage caused by leaks.
13. All work shall comply with NFPA standards as applicable.
14. All work shall comply with all local rules and regulations as defined by the authority having jurisdiction.
15. All work shall comply with OSHA regulations.

Appendix D: Portable and Wheeled Fire-Extinguisher Guide

(Courtesy, Amerex Corporation)

WHERE TO USE

It is natural for a person to use the extinguisher located nearest to a fire, making it essential that the correct type and size be placed in close proximity to a potential hazard. The most current issue of NFPA-10 should be consulted for minimum recommended fire-extinguisher types, placement, and travel distances.

All fire-extinguisher nameplates have either the letter or picture symbols shown below. Anyone who might be expected to use a fire extinguisher should be familiar with the letter or picture symbols identifying the type(s) of fire on which it may be used. The newer picture symbols use the international sign system diagonal red slash to indicate a potential hazard if the extinguisher is used on that particular type of fire. Absence of a type of symbol means only that the unit is not recommended as particularly effective for that class, not that it is dangerous if used in error.

TYPES OF FIRES

LETTER SYMBOL		PICTURE SYMBOL
	For wood, paper, cloth, trash, and other ordinary materials.	
	For gasoline, grease, oil, paint, and other flammable liquids.	
	For live electrical equipment.	
	For combustible metals.	**No Current Symbol**

TYPES OF EXTINGUISHERS

CLASS A

CLASS AB

CLASS BC

CLASS ABC

HOW TO USE

All new extinguishers are furnished with a detailed owner's manual containing valuable information. It and the extinguisher nameplate contain the "How To Use" illustrations shown below. Potential operators should be totally familiar with these instructions.

(1)	(2)	(3)
HOLD UPRIGHT. PULL RING PIN.	START BACK 10 FEET. AIM AT BASE OF FIRE.	SQUEEZE LEVER. SWEEP SIDE TO SIDE.

ALL FIRE EXTINGUISHERS SHOULD COMPLY WITH THE RECOMMENDATIONS OF THE NATIONAL FIRE PROTECTION ASSOCIATION AND BE TESTED AND RATED BY UNDERWRITERS LABORATORIES OR FACTORY MUTUAL SYSTEMS TO ANSI/UL SPECIFICATIONS. TO ASSIST IN COMPLYING WITH NATIONAL AND LOCAL OSHA REQUIREMENTS, ALL NAMEPLATES SHOULD CONTAIN REQUIRED HMIS INFORMATION.

A wide variety of types and sizes of hand portable and wheeled fire extinguishers are manufactured. Due consideration should be given to the type of hazard and potential size of fire involvement. Your distributor is in a position to advise on size, type, and location of equipment. Caution is urged in the placement of dry chemical extinguishers where there is potential for damage to delicate electrical equipment from the extinguishing agent. ABC dry chemical derives its Class A firefighting capabilities from the fact that it melts at approximately 350°F, clings to the surface to which it is applied, and smothers the fire. This effect may dictate the selection of other agents such as Regular or Purple-K dry chemical, Halon 1211 or CO_2 for specific applications. Halon 1211 leaves no residue and is generally recommended for placement in areas where delicate electrical (computers, generators, switch panels, etc.) or aircraft might be involved.

The number and types of fire extinguishers are normally delermined by THE LOCAL AUTHORITIES HAVING JURISDICTION. Many local regulations are taken from the recommendations of National Fire Protection Standard no. 10, titled "Standard For Portable Fire Extinguishers." Pertinent information from NFPA-10 is listed below.

Extinguisher Selection

C-1   Principles of Selecting Extinguishers.

C-1-1   Selection of the best portable fire extinguisher for a given situation depends on:

(a) the nature of the combustibles that might be ignited
(b) the potential severity (size, intensity, and speed of travel) of any resulting

fire

    (c) effectiveness of the extinguisher on that hazard

    (d) the ease of use of the extinguisher

    (e) the personnel available to operate the extinguisher and their physical abilities and emotional reactions as influenced by their training

    (f) the ambient temperature conditions and other special atmospheric considerations (wind, draft, presence of fumes)

    (g) suitability of the extinguisher for its environment

    (h) any anticipated adverse chemical reactions between the extinguishing agent and the burning materials

    (i) any health and operational safety concerns (exposure of operators during the fire-control efforts), and

    (j) the upkeep and maintenance requirements for the extinguisher.

C-3-7 Wheeled Extinguishers.

C-3-7.1 The selection of any type of wheeled extinguisher is generally associated with a recognized need to provide additional protection for special hazards or large, extra-hazard areas. Where wheeled extinguishers are to be installed, consideration should be given to mobility within the area in which they will be used.

C-3-7.2 For outdoor locations, models with rubber tires or wide-rim wheels will be easier to transport. For indoor locations, doorways, aisles, and corridors need to be wide enough to permit the ready passage of the extinguisher. Because of the magnitude of the fire it will generally be used on, this type of extinguisher should be reserved for use by operators who have actually used the equipment, who have received special instructions on the use of the equipment, or who have used the equipment in live fire training

1-5 Classification of Hazards.

1-5.1 Light (Low) Hazard. Locations where the total amount of Class A combustible materials, including furnishings, decorations and contents, is minor. These may include buildings or rooms occupied as offices, classrooms, churches, assembly halls, etc. This classification anticipates that the majority of contents items are either noncombustible or so arranged that a tire is not likely to spread rapidly. Small amounts of Class B flammables used for duplicating machines, art departments, etc., are included provided that they are kept in closed containers and safely stored.

1-5.2 Ordinary (Moderate) Hazard. Locations where the total amount of Class A combustibles and Class B flammables are present in greater amounts than expected under Light (Low) Hazard occupancies. These occupancies could consist of offices, classrooms, mercantile shops and allied storage, light manufacturing, research operations, auto showrooms, parking garages, workshop or support service areas of Light (Low) Hazard occupancies and warehouses containing Class I or Class 11 commodities as defined by NFPA 231, *Standard for Indoor General Storage.*

1-5.3 Extra (High) Hazard. Locations where the total amount of Class A combustibles and Class B flammables are present, in storage, production use and/or finished product over and above those expected and classed as ordinary (moderate) hazards. These occupancies could consist of woodworking, vehicle repair, aircraft and boat servicing, individual product display showrooms, product convention center displays storage and manufacturing processes such as painting, dipping, coating, including flammable liquid handling. Also included is warehousing of, or in-process storage of other than Class I and Class 11 commodities.

2-2 Selection by Hazard.

2-2.1   Extinguishers shall be selected for the specific class or classes of hazards to be protected in accordance with the following subdivisions.

2-2.1.1 Extinguishers for protecting Class A hazards shall be selected from the following: water, antifreeze protected water, aqueous film forming foam (AFFF), film forming fluoroprotein (FFFP), welling agent, loaded stream, and multipurpose dry chemical.

2-2.1.2 Extinguishers for protection of Class B hazards shall be selected from the following: carbon dioxide, dry chemical types, aqueous film forming foam (AFFF) and film forming fluoroprotein (FFFP).

2-2.1.3 Extinguishers for protection of Class C hazards shall be selected from the following: carbon dioxide, dry chemical types.'

2-2.1.4 Extinguishers and extinguishing agents for the protection of Class D hazards shall be of types approved for use on the specific combustible-metal hazard.

3-2   **Fire Extinguisher Size and Placement for Class A Hazards.**

Table 3-2.1

	Light (Low) Hazard Occupancy	Ordinary (Moderate) Hazard Occupancy	Extra (High) Hazard Occupancy
Minimum rated single extinguisher	2-A	2-A	4-A*
Maximum floor area per unit of A	3,000 sq ft	1,500 sq ft	1,000 sq ft
Maximum floor area for extinguisher	11,250 sq ft	1,250 sq ft	11,250 sq ft
Maximum travel distance to extinguisher	75 ft	75 ft	75 ft

* Two 2½-gal (9.46 L) water type extinguishers can be used to fulfill the requirements of one 4-A rated extinguisher.

NOTE: 1 ft = 0.305 m
      1 sq ft = 0.0929 m^2

3-2.1.1 Certain smaller extinguishers which are charged with multipurpose dry chemical or Halon 1211 are rated on Class B and Class C fires, but have insufficient effectiveness to earn the minimum 1-A rating even though they have value in extinguishing smaller Class A fires. They shall not be used to meet the requirements of 3-2.1.

3-2.2   Up to one-half of the complement of extinguishers as specified in Table 3-2.1 may be replaced by uniformly spaced 1½-in. (3.81-cm) hose stations for use by the occupants of the building. When hose stations are so provided they shall conform to NFPA 14, Installation of Standpipe and Hose Systems. The location of hose stations and the placement of fire extinguishers shall be in such a manner that the hose stations do not replace more than every other extinguisher.

3-3   **Fire Extinguisher Size and Placement for Class B Fires Other than for Fires in Flammable Liquids of Appreciable Depth.**

3-3.1   Minimal sizes of fire extinguishers for the listed grades of hazard shall be provided on the basis of Table 3-3.1. Extinguishers shall be located so that the maximum travel distances shall not exceed those specified in the table used.

Exception: Extinguishers of lesser rating, desired for small specific hazards within the general hazard area, may be used, but shall not be considered as fulfilling any part of the requirements of Table 3-3.1.

Table 3-3.1

Type of Hazard	Basic Minimum Extinguisher Rating	Maximum Travel Distance to Extinguishers, Ft. (m)
Light (low)	5B	30 (9.15)
	10B	50 (15.25)
Ordinary (moderate)	10B	30 (9.15)
	20B	50 (15.25)
Extra (high)	40B	30 (9.15)
	80B	50 (15.25)

NOTE: The specified ratings do not imply that fires of the magnitudes indicated by these ratings will occur, but rather are to give the operators more time and agent to handle difficult spill fires that may occur.

3-3.2   Two or more extinguishers of lower rating shall not be used to fulfill the protection requirements of Table 3-3.1.

Exception No. 1: Up to three foam extinguishers of at least 2½ gal (9.46 L) capacity may be used to fulfill light (low) hazard requirements.

Exception No. 2: Up to three AFFF or FFFP extinguishers of at least 2½ gal (9.46 L) capacity may be used to fulfill extra (high) hazard requirements.

3-3.3   The protection requirements may be fulfilled with extinguishers of higher ratings provided the travel distance to such larger extinguishers shall not exceed 50 ft (15.25 m).

3-4   Fire Extinguisher Size and Placement for Class B Fires in Flammable Liquids of Appreciable Depth.[1]

3-4.1   Portable fire extinguishers shall not be installed as the sole protection for flammable liquid hazards of appreciable depth (greater than ¼ in. [0.64 cm]) where the surface area exceeds 10 sq ft (0.93 sq m).

Exception: Where personnel who are trained in extinguishing fires in the protected hazards, or a counterpart, are available on the premises, the maximum surface area shall not exceed 20 sq ft (1.86 sq m).

3-4.2   For flammable liquid hazards of appreciable depth such as in dip or quench tanks, a Class B fire extinguisher shall be provided on the basis of at least two numerical units of Class B extinguishing potential per sq ft (0.0929 sq m) of flammable liquid surface of the largest tank hazard within the area.

Exception No. 1: Where approved automatic fire protection devices or systems have been installed for a flammable liquid hazard, additional portable Class B fire extinguishers may be waived. Where so waived, Class B extinguishers shall be provided as covered in 3-3.1 to protect areas in the vicinity of such protected hazards.

Exception No. 2: AFFF or FFFP type extinguishers may be provided on the basis of IB of protection per sq ft of hazard.

---

[1] For dip tanks containing flammable or combustible liquids exceeding 150 gal (568 L) liquid capacity or having a liquid surface exceeding 4 sq ft (0.38 sq m), see NFPA34, *Dip Tanks*, for requirements of automatic extinguishing facilities.

References: ANSI / NFPA 10, ANSI / NFPA 96

Available from: National Fire Protection Association, Inc.
                                   Batterymarch Park
                                   Quincy, MA 02269

Application for Specific Hazards.

Class B Fire Extinguishers for Pressurized Flammable Liquids and Pressurized Gas Fires. Fires of this nature are considered to be a special hazard. Class B fire extinguishers containing agents other than dry chemical are relatively ineffective on this type of hazard due to stream and agent characteristics. Selection of extinguishers for this type of hazard shall be made on the basis of recommendations by manufacturers of this specialized equipment. The system used to rate extinguishers on Class B fires (flammable

HAND PORTABLE & WHEEL

TYPES OF FIRE	TYPES OF HAZARDS	RECOMMENDED TYPE EXTINGUISHER	EFFECTIVE EXTINGUISHING ACTION
CLASS A	ORDINARY COMBUSTIBLES	Water	Quench, Cool, Penetrate
	Wood, Paper, Fabrics, Rubber, Many Plastics	AFFF (FFFP)	Quench, Cool, Penetrate
		ABC Dry Chemical	Chemical reaction, Coat, Cool
		Halon 1211	Chemical reaction, Quench, Cool
CLASS B	FLAMMABLE LIQUIDS AND GASES	*Liquids:* AFFF (FFFP)	Smother, Cool
		Regular Dry Chemical	Chemical Reaction, Smother
		ABC Dry Chemical	Chemical Reaction, Smother
	Gasoline, Paint, Oils,	Purple K Dry Chemical	Chemical Reaction, Smother
	Lacquer, Grease, Tar,	Halon 1211	Chemical Reaction, Smother, Cool
	Natural & Manufactured	CO_2	Smother, Cool
	Gases	*Gases:* Regular & Purple K Dry Chemical	Smothering Action Preferred
CLASS C	ENERGIZED ELECTRICAL EQUIPMENT	Regular Dry Chemical ABC Dry Chemical	All Are Non-
	Wiring, Motors, Switches,	Purple K Dry Chemical	Conductors of Electricity
	Generators, Panels, Appliances	Halon 1211 CO_2	
CLASS AB	ORDINARY COMBUSTIBLES FLAMMABLE LIQUIDS & GASES	AFFF (FFFP) ABC Dry Chemical	Smother, Cool Chemical Reaction, Smother
	Combinations of A & B	Halon 1211	Chemical Reaction, Smother, Cool
CLASS BC	FLAMMABLE LIQUIDS & GASES	Regular Dry Chemical	Chemical Reaction, Smother, Non-conductor
		ABC Dry Chemical	Chemical Reaction, Smother, Non-conductor
	ENERGIZED ELECTRICAL	Purple K Dry Chemical	Chemical Reaction, Smother, Non-conductor
	EQUIPMENT	Halon 1211	Chemical Reaction, Smother, Cool, Non-condu
	Combinations of B & C	CO_2	Smother, Non-conductor
CLASS ABC	ORDINARY COMBUSTIBLES FLAMMABLE LIQUIDS & GASES ENERGIZED ELECTRICAL EQUIPMENT	ABC Dry Chemical	Chemical Reaction, Smother, Non-conductor
	Combinations of A, B & C	Halon 1211	Chemical Reaction, Quench, Smother, Cool, Non-conductor
CLASS D	COMBUSTIBLE METALS & COMBUSTIBLE METAL ALLOYS	Super D (Sodium Chloride) G-Plus (Graphite)	Excludes Air, Dissipates Heat Smother, Absorbs Heat

liquids in depth) is not applicable to these types of hazards. It has been determined that special nozzle design and rates of agent application are required to cope with such hazards. Caution: It is undesirable to attempt to extinguish this type of fire unless there is reasonable assurance that the source of fuel can be promptly shut off.

Cooking Grease Fires. Fires involving liquefied fat or oil in depth, such as fat fryers, are considered to be a special hazard. Only extinguishers containing sodium bicarbonate or potassium bicarbonate dry chemicals have been proven effective on this hazard due to agent characteristics.

RE EXTINGUISHER GUIDE

EFFECTIVE OPERATING TEMPERATURE RANGE (°F)	APPROXIMATE DISCHARGE TIME (SEC)		APPROXIMATE DISCHARGE RANGES (FT)		AVAILABLE SIZES	
	Hand Portable	Wheeled	Hand Portable	Wheeled	Hand Portable	Wheeled
+40 to +120 (May be freeze-protected to –40)	45 – 50	N/A	30 – 45	N/A	2½ gal.	N/A
+40 to +120	60 – 65	55 – 65	15 – 20	30 – 40	2½ gal.	33 gal.
–65 to +120	10 – 30	48 – 52	9 – 21	30 – 40	2½,5,6,10,20 lb.	50 & 125 lb.
–65 to +120	10 – 23	44	9 – 18	30 – 40	1¼,2½,5,9,13,17,20 lb.	50 & 150 lb.
+40 to +120	60 – 65	55 – 65	15 – 20	30 – 40	2½ gal.	33 gal.
–65 to –120	10 – 30	53 – 60	9 – 21	30 – 40	2½,5,5½,6,10,20 lb.	50 & 150 lb.
–65 to +120	10 – 30	48 – 52	9 – 21	30 – 40	2½,5,6,10,20 lb.	50 & 125 lb.
–65 to +120	10 – 30	52 – 53	9 – 21	30 – 40	2½,5,10,20 lb.	50 & 125 lb.
–65 to +120	10 – 23	44	9 – 18	30 – 40	1¼,2½,5,9,13,17,20 lb.	50 & 150 lb.
–40 to –120	10 – 20	44 – 76	3 – 8	10 – 15	5,10,15,20 lb.	50 & 100 lb.
–65 to +120	10 – 30	52 – 60	9 – 21	30 – 40	2½,5,6,10,20 lb.	50 & 125 lb.
–65 to +120	10 – 30	53 – 60	9 – 21	30 – 40	2½,5,5½,6,10,20 lb.	50 & 150 lb.
–65 to +120	10 – 30	48 – 52	9 – 21	30 – 40	2½,5,6,10,20 lb.	50 & 125 lb.
–65 to +120	10 – 30	52 – 53	9 – 21	30 – 40	2½,5,10,20 lb.	50 & 125 lb.
–65 to +120	10 – 23	44	9 – 18	30 - 40	1¼,2½,5,9,13,17,20 lb.	50 & 150 lb.
–40 to +120	10 – 20	44 – 74	3 – 8	10 – 15	5,10,15,20 lb.	50 & 100 lb.
+40 to +120	60 – 65	55 – 65	30 – 45	30 – 40	2½ gal.	33 gal.
–65 to +120	10 – 30	48 – 52	9 – 21	30 – 40	2½,5,6,10,20 lb.	50 & 125 lb.
–65 to +120	10 – 23	44	9 – 18	30 – 40	1¼,2½,5,9,13,17,20 lb.	50 & 150 lb.
–65 to +120	10 – 30	53 – 60	9 – 21	30 – 40	2½,5,5½,6,10,20 lb.	50 & 150 lb.
–65 to +120	10 – 30	48 – 52	9 – 21	30 – 40	2½,5,6,10,20 lb.	50 & 125 lb.
–65 to +120	10 – 30	52 – 53	9 – 21	30 – 40	2½,5,10,20 lb.	50 & 125 lb.
–65 to +120	10 – 23	44	9 – 18	30 – 40	1¼,2½,5,9,13,17,20 lb.	50 & 150 lb.
–40 to +120	10 – 20	44 – 74	3 – 8	10 – 15	5,10,15,20 lb.	50 & 100 lb.
–65 to +120	10 – 30	48 – 52	9 – 21	30 – 40	2½,5,6,10,20 lb.	50 & 125 lb.
–65 to +120	10 – 23	44	9 – 18	30 – 40	1¼,2½,5,9,13,17,20 lb.	50 & 150 lb.
–40 to +120	28 – 30	N/A	4 – 6	N/A	30 lb.	N/A
N/A	N/A	N/A	N/A	N/A	40 lb. Pail (use scoop or shovel)	N/A

NOTE: NFPA 96 "Vapor Removal From Cooking Equipment 1984" now states that portable extinguishers shall be installed in kitchen cooking areas in accordance with NFPA 10 Table 3-3.1 for Extra (High) Hazard.

Three-Dimensional Class B Fires. A three-dimensional Class B fire involves Class B materials in motion such as pouring, running, or dripping flammable liquids and generally includes vertical as well as one or more horizontal surfaces. Fires of this nature are considered to be a special hazard. Selection of extinguishers for this type of hazard shall be made on the basis of recommendations by manufacturers of this specialized equipment. The system used to rate extinguishers on Class B fires (flammable liquids in depth) is not directly applicable to this type of hazard.

NOTE: The installation of fixed systems should be considered when applicable.

Class B Fire Involving Polar or Water Soluble Flammable Liquids. Film forming fluoroprotein (FFFP) and certain grades of Aqueous film forming foam (AFFF) are suitable for the protection of water-soluble flammable liquids (polar solvents) such as alcohols, acetone, esters, ketones, etc. The suitability of these extinguishers for polar solvent fires must be specifically referenced on the extinguisher nameplate.

Electronic Equipment Fires. Extinguishers for the protection of delicate electronic equipment shall be selected from the following: clean gas (Inergen) and carbon dioxide.

Appendix E: Sample of a Computer Hydraulic Calculation

PROTECTED AREA

Company Name _____

Address   _____

Telephone No. _____

Date: 1992

Contractor: Fire Protection Co., Inc.

Property: Store No._____   High Area Using Loop

The design criteria for this set of calculations are 0.15 gpm/sq ft for the most remote 2500 square feet including 250 gpm for inside hose and 0 gpm for outside hydrants.

The total final water demand is 922.2 gpm at 55.4 psi.

The available water supply is 922.2 gpm at 56.0 psi.

Water supply data:

Flow test static pressure 163 psi, residual pressure 129 psi, flow 1000 gpm. The elevation of the water test is the same as the lowest floor.

General comments: None

Sprklr. Ident.	Total Heads	Q(A) Flow Q(T)	Dia	Pipe & Equivalent Lengths		Loss psi/ft	Friction PF PE	PT Notes
8	1	0.0	1.0490	Lgth	6	0.0746	7.0	
				Ftg	5		0.8	Vel = 5.5 fps
		14.8		Total	11		0.0	
9	2	15.1	1.0490	Lgth	7	0.2738	7.8	
				Ftg	0		1.9	Vel = 11.1 fps
		29.9		Total	7		0.0	
10	3	16.8	1.3800	Lgth	4	0.1643	9.7	
				Ftg	6		1.6	Vel = 10.0 fps
		46.7		Total	10		0.0	
15							11.3	K = 13.9
11	1	0.0	1.0490	Lgth	10	0.0746	7.0	
				Ftg	0		0.7	Vel = 5.5 fps
		14.8		Total	10		0.0	
12	2	15.5	1.0490	Lgth	12	0.2806	7.7	
				Ftg	8		5.6	Vel = 11.3 fps
		30.3		Total	20		0.0	
13	3	20.4	1.3800	Lgth	2	0.1913	13.3	
				Ftg	0		0.4	Vel = 10.9 fps
		50.7		Total	2		0.0	
42	4	19.6	1.6100	Lgth	4	0.1653	13.7	
				Ftg	8		2.0	Vel = 11.1 fps
		70.3		Total	12		0.0	
14	5	21.0	1.6100	Lgth	6	0.2681	15.7	
				Ftg	0		1.6	Vel = 14.4 fps
		91.3		Total	6		0.0	
15	8	57.8	2.0680	Lgth	3	0.1963	17.3	
				Ftg	10		2.6	Vel = 14.3 fps
		149.1		Total	13		0.0	
16	14	123.1	2.6350	Lgth	8	0.1837	19.9	
				Ftg	18		4.8	Vel = 16.0 fps
		272.2		Total	26		−2.6	

Sprklr. Ident.	Total Heads	Q(A) Flow Q(T)	Dia	Pipe & Equivalent Lengths		Loss psi/ft	Friction PF PE	PT Notes
A							22.1	K = 57.9
		0.0		Lgth	6		7.0	
17	1		1.0490	Ftg	2	0.0746	0.8	Vel = 5.5 fps
		14.8		Total	8		0.0	
		14.6		Lgth	5		7.6	
18	2		1.0490	Ftg	0	0.2654	1.3	Vel = 10.9 fps
		29.4		Total	5		0.0	
		16.1		Lgth	2		8.9	
19	3		1.3800	Ftg	6	0.1566	1.3	Vel = 9.8 fps
		45.5		Total	8		0.0	
36B							10.2	K = 14.2
		0.0		Lgth	8		7.0	
20	1		1.0490	Ftg	2	0.0746	0.7	Vel = 5.5 fps
		14.8		Total	10		0.0	
		15.0		Lgth	3		7.7	
21	2		1.0490	Ftg	5	0.2721	2.2	Vel = 11.2 fps
		30.1		Total	8		0.0	
43							9.9	K = 9.5
		0.0		Lgth	10		7.0	
22	1		1.0490	Ftg	4	0.0746	1.0	Vel = 5.5 fps
		14.8		Total	14		0.0	
		15.3		Lgth	3		8.0	
23	2		1.0490	Ftg	5	0.2772	2.2	Vel = 11.2 fps
		30.1		Total	8		0.0	
		17.2		Lgth	8		10.2	
24	3		1.3800	Ftg	0	0.1682	1.3	Vel = 10.2 fps
		47.3		Total	8		0.0	
		32.2		Lgth	1		11.5	
43	5		1.6100	Ftg	8	0.2075	1.9	Vel = 12.5 fps
		79.5		Total	9		0.0	

Sprklr. Ident.	Total Heads	Q(A) Flow Q(T)	Dia	Pipe & Equivalent Lengths		Loss psi/ft	Friction PF PE	PT Notes
		52.0		Lgth	6		13.4	
36B	8		2.0680	Ftg	10	0.1556	2.5	Vel = 12.6 fps
		131.5		Total	16		−2.6	
36T							13.3	K = 36.1
		0.0		Lgth	8		7.0	
25	1		1.0490	Ftg	0	0.0746	0.6	Vel = 5.5 fps
		14.8		Total	8		0.0	
		15.4		Lgth	8		7.6	
26	2		1.0490	Ftg	0	0.2789	2.2	Vel = 11.2 fps
		30.2		Total	8		0.0	
		17.5		Lgth	8		9.8	
27	3		1.3800	Ftg	9	0.1709	2.9	Vel = 10.2 fps
		47.7		Total	17		0.0	
35							12.7	K = 13.4
		0.0		Lgth	6		7.0	
28	1		1.0490	Ftg	0	0.0746	0.4	Vel = 5.5 fps
		14.8		Total	6		0.0	
		14.7		Lgth	11		7.4	
29	2		1.0490	Ftg	0	0.2671	2.9	Vel = 11.0 fps
		29.5		Total	11		0.0	
		17.3		Lgth	5		10.3	
30	3		1.3800	Ftg	9	0.1650	2.3	Vel = 10.0 fps
		46.8		Total	14		0.0	
34							12.6	K = 13.2
		0.0		Lgth	8		7.2	
31	1		1.0490	Ftg	0	0.0764	0.6	Vel = 5.6 fps
		15.0		Total	8		0.0	
		15.6		Lgth	8		7.8	
32	2		1.0490	Ftg	0	0.2858	2.3	Vel = 11.4 fps
		30.6		Total	8		0.0	

Sprklr. Ident.	Total Heads	Q(A) Flow Q(T)	Dia	Pipe & Equivalent Lengths		Loss psi/ft	Friction PF PE	PT Notes
		17.8		Lgth	12		10.1	
33	3		1.3800	Ftg	3	0.1756	2.6	Vel = 10.4 fps
		48.4		Total	15		0.0	
		47.0		Lgth	4		12.7	
34	6		2.0680	Ftg	10	0.0859	1.2	Vel = 9.1 fps
		95.4		Total	14		0.0	
		50.0		Lgth	9		13.9	
35	9		2.0680	Ftg	0	0.1874	1.7	Vel = 13.9 fps
		145.4		Total	9		0.0	
		142.6		Lgth	3		15.6	
36	17		2.6350	Ftg	12	0.2039	3.1	Vel = 17.0 fps
		288.0		Total	15		0.0	
B							18.7	K = 66.6
		0.0		Lgth	14		7.0	
37	1		1.3800	Ftg	0	0.0196	0.3	Vel = 3.2 fps
		14.8		Total	14		0.0	
		14.6		Lgth	3		7.3	
38	2		1.6100	Ftg	8	0.0329	0.4	Vel = 4.6 fps
		29.4		Total	11		0.0	
		14.7		Lgth	9		7.7	
39	3		1.6100	Ftg	4	0.0698	0.9	Vel = 7.0 fps
		44.1		Total	13		0.0	
		0.0		Lgth	9		8.6	
40	3		2.0680	Ftg	10	0.0206	0.4	Vel = 4.2 fps
		44.1		Total	19		0.0	
		0.0		Lgth	8		9.0	
41	3		3.2600	Ftg	22	0.0022	0.1	Vel = 1.7 fps
		44.1		Total	30		−2.6	
C							6.5	K = 17.3

Sprklr. Ident.	Total Heads	Q(A) Flow Q(T)	Dia	Pipe & Equivalent Lengths	Loss psi/ft	Friction PF PE	PT Notes
This is for the clockwise flow:							
		0.0		Lgth 145		22.1	
A			3.2600	Ftg 44	0.0642	12.1	Vel = 10.4 fps
		270.0		Total 189		0.0	
E						34.2	K = 46.2
This is for the counterclockwise flow:							
		0.0		Lgth 32		22.1	
A			3.2600	Ftg 0	0.0000	0.0	Vel = 0.1 fps
		2.8		Total 32		0.0	
		313.1		Lgth 33		22.1	
B			3.2600	Ftg 0	0.0858	2.8	Vel = 12.2 fps
		315.9		Total 33		0.0	
		86.3		Lgth 47		24.9	
C			3.2600	Ftg 22	0.1341	9.3	Vel = 15.5 fps
		402.2		Total 69		0.0	
D						34.2	K = 68.8
		0.0		Lgth 20		34.2	
E	34		3.2600	Ftg 41	0.3469	21.2	Vel = 25.9 fps
		672.2		Total 61		0.0	
		250.0		Lgth 0		55.4	
Z	34		3.2600	Ftg 0	0.6226	0.0	Vel = 35.5 fps
		922.2		Total 0		0.0	
Z						55.4	

The water demand is 922.2 gpm at 55.4 psi.

The water supply available is 922.2 gpm at 56.0 psi.

Appendix F: Sprinkler-Room Arrangement, Backflow-Preventer Detail

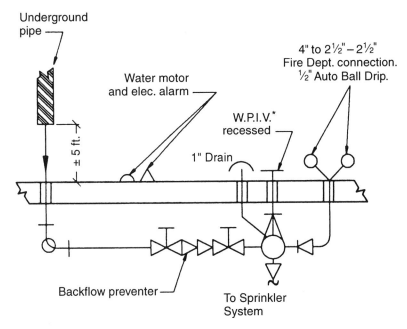

NOTE: An approximate room size for a system requiring an 8" riser is 8 ft × 16 ft.

* W.P.I.V. = Wall Post Indicator Valve

**SPRINKLER ROOM
WATER SUPPLY ARRANGEMENT
PLAN DETAIL**
(No Scale)

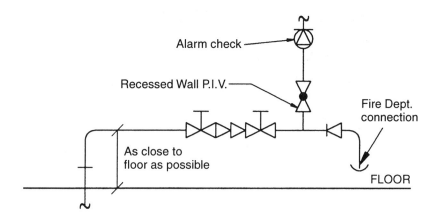

SPRINKLER ROOM RISER DIAGRAM
(No Scale)

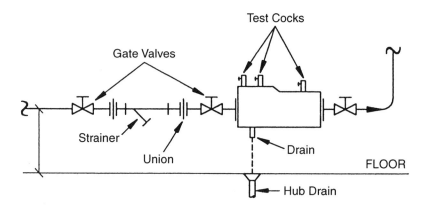

BACKFLOW PREVENTER SCHEMATIC
(Not to Scale)

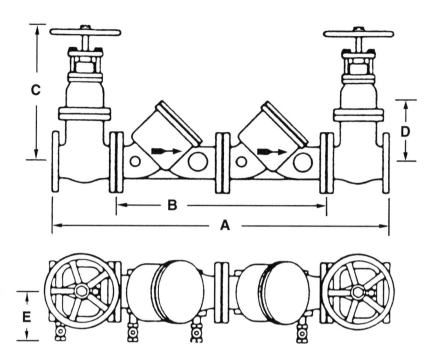

BACKFLOW PREVENTER (BFP)
SIZES AND DIMENSIONS*

Size	Dimension (in.)					Weight
(in.)	A	B	C	D	E	(lb)
$2^1/_2$	$37^3/_{16}$	$22^1/_{16}$	$12^1/_2$	$7^1/_2$	$5^1/_4$	240
3	$41^{11}/_{16}$	$25^9/_{16}$	14	$8^1/_{16}$	6	250
4	$50^7/_{16}$	$32^5/_{16}$	$17^3/_8$	11	$6^3/_4$	420
6	$59^{11}/_{16}$	$38^9/_{16}$	$21^1/_4$	14	$8^1/_4$	735
8	$69^3/_{16}$	$46^1/_{16}$	26	18	$9^1/_2$	1230
10	$84^3/_{16}$	$58^1/_{16}$	30	22	$10^1/_2$	1680

* For information only. A BFP is used to prevent backflow of pollutants. It must not be used to isolate toxic fluids.

Appendix G: Extinguisher Selection Chart

Quick Selector Chart

Find your special hazard listed below. The most commonly used extinguishing agents are checked. If for any reason these agents are undesirable at your location, others not checked may be used. If your hazard is not listed, consult the applicable NFPA Standard or manufacturer recommendation.

Area to Be Protected	Type of Protection				
Special Fire Hazard	**Water Spray**	**Foam**	**Carbon Dioxide**	**Dry Chemical**	**Halon**
Aircraft Hangars	•	•			
Alcohol Storage	•	•	•		
Ammunition Loading	•				
Ammunition Magazines	•				
Asphalt Impregnating	•				
Battery Rooms			•		
Carburetor Overhaul Shops	•	•	•	•	
Cleaning Plant Equipment	•	•	•	•	
Computer Rooms					•
Dowtherm	•				
Drying Ovens	•		•	•	
Engine Test Cells	•	•	•		
Escalators, Stairwells	•				
Explosives: Manufacturing, Storage	•				
Flammable Liquids Storage	•	•	•		
Flammable Solids Storage	•				
Fuel Oil Storage	•	•			
Hangar Decks	•	•			
Hydraulic Oil, Lubricating Oil	•		•		
Hydro-Turbine Generators	•		•		•
Jet Engine Test Cells	•	•	•		•
Lignite Storage and Handling	•				
Liquefied Petroleum Gas Storage	•				
Oil Quenching Bath	•	•	•	•	
Paints: Manufacturing, Storage	•	•	•	•	•
Paint Spray Booths	•		•	•	
Petrochemical Storage	•	•	•		•
Petrochemical Testing Laboratories	•	•	•		•
Printing Presses			•		
Reactor and Fractionating Towers	•				
Record Vaults			•		•
Rubber Mixing and Heat Treating	•				
Shipboard Storage	•		•		•
Solvent Cleaning Tanks		•	•	•	
Solvent Thinned Coatings		•	•	•	
Switchgear Rooms			•		•
Transformer, Circuit Breakers (Outdoors)	•				
Transformer, Circuit Breakers (Indoors)	•		•		
Turbine Lubricating Oil	•	•	•	•	
Vegetable Oil, Solvent Extraction	•	•			

Source: Courtesy, Grinnell Corporation.

Appendix H: Unit and Conversion Factors

To convert from other systems of measurement to SI values, the following conversion factors are to be used. (Note: For conversion equivalents not shown herein, refer to ANSI Z210.1, also issued as ASTM E380.)

Unit and Conversion Factors

a. **Linear acceleration**

ft/s^2 = 0.3048 m/s^2 m/s^2 = 3.28 ft/s^2
$in./s^2$ = 0.0254 m/s^2 m/s^2 = 9.37 $in./s^2$

b. **Area**

acre = 4046.9 m^2 m^2 = 0.0000247 acre
ft^2 = 0.0929 m^2 m^2 = 10.76 ft^2
$in.^2$ = 0.000645 m^2 = 645.16 mm^2 m^2 = 1550.39 in^2
mi^2 = 2 589 988 m^2 km^2 = 0.39 mi^2
yd^2 = 0.836 m^2 m^2 = 1.2 yd^2

c. **Bending moment (torque)**

pound–force·inch (lbf·in.) =
    0.113 Newton meter (N·m)   N·m = 8.85 lbf·in.
lbf·ft = 1.356 N·m   N·m = 0.74 lbf·in.

d. **Bending moment (torque) per unit length**

lbf·in./in. = 4.448 N·m/m   N·m/m = 0.225 lbf·in./in.
lbf·ft/in. = 53.379 N·m/m   N·m/m = 0.019 lbf·ft/in.

e. **Electricity and magnetism (designation only)**

ampere = 1 A
ampere–hour = 3600 Ah
coulomb = 1 C
farad = 1 F
henry = 1 H
ohm = 1 Ω
volt = 1 V

f. Energy (work)

British thermal unit (Btu) =
  1055 Joule (J)   J = 0.000948 Btu
ft·lbf = 1.356 J   J = 0.074 ft·lbf
kWh = 3 600 000 J   J = 0.000000278 kWh

g. Energy per unit area per unit time

Btu/(ft^2·s) = 11 349 W/m^2 W/m^2 = 0.000088 Btu/(ft^2·s)

h. Force

ounce–force (ozf) = 0.287 N   N = 3.48 ozf
pound–force (lbf) = 4.448 N   N = 0.23 lbf
kilogram–force (kgf) = 9.807 N   N = 0.1 kgf

i. Force per unit length

lbf/in. = 175.1 N/m   N/m = 0.0057 lbf/in.
lbf/ft = 14.594 N/m   N/m = 0.069 lbf/ft

j. Host

Btu·in./(s·ft^2·°F) = 519.2 W/(m·K)   W/(m·K) = 0.002 Btu·in./(s·ft^2)
Btu·in./(h·ft^2·°F) = 0.144 W/(m·K)   W/(m·K) = 6.94 Btu·in./(h·ft^2)
Btu/ft^2 = 11 357 J/m^2 J/m^2 = 0.000088 Btu/ft^2
Btu/(h·ft^2·°F) = 5.678 W/(m^2·K)   W/(m^2·K) = 0.176 Btu/(h·ft^2·°F)
Btu/lbm = 2326 J/kg   J/kg = 0.00043 Btu/lbm
Btu/(lbm·°F) = 4186.8 J/(kg·K)   J/(kg·K) = 0.000239 Btu/(lbm·°F)
(°F·h·ft^2)/Btu = 0.176 (K·m^2)/W   (K·m^2)/W = 5.68 (°F·h·ft^2)/Btu

k. Length

in. = 0.0254 m   m = 39.37 in.
ft = 0.3048 m   m = 3.28 ft
yd = 0.914 m   m = 1.1 yd
mi = 1609.3 m   m = 0.000621 mi

l. Light (illuminance)

footcandle (fc) = 10.764 lx   lx = 0.093 fc

m. Mass

ounce–mass (ozm) = 0.028 kg   kg = 35.7 ozm
pound–mass (lbm) = 0.454 kg   kg = 2.2 lbm

n. Mass per unit area

lbm/ft^2 = 4.882 kg/m^2 kg/m^2 = 0.205 lbm/ft^2

o. Mass per unit length

lbm/ft = 1.488 kg/m   kg/m = 0.67 lbm/ft

p. Mass per unit time (flow)

lbm/h = 0.0076 kg/s   kg/s = 131.58 lbm/h

q. Mass per unit volume (density)

$lbm/ft^3 = 16.019 \ kg/m^3$ $kg/m^3 = 0.062 \ lbm/ft^3$
$lbm/in^3 = 27,680 \ kg/m^3$ $kg/m^3 = 0.000036 \ lbm/in^3$
$lbm/gal = 119.8 \ kg/m^3$ $kg/m^3 = 0.008347 \ lbm/gal$

r. Moment of inertia

$lb/ft^2 = 0.042 \ kg \cdot m^2$ $kg \cdot m^2 = 23.8 \ lb/ft^2$

s. Plane angle

degree = 17.453 mrad   mrad = 0.057 deg
minute = 290.89 μrad   μrad = 0.00344 min
second = 4.848 μrad   μrad = 0.206 s

t. Power

Btu/h = 0.293 W   W = 3.41 Btu/h
(ft·lbf)/h = 0.38 mW   mW = 2.63 (ft·lbf)/h
horsepower (hp) = 745.7 W   W = 0.00134 hp

u. Pressure (stress) force per unit area

atmosphere = 101.325 kiloPascal (kPa)   kPa = 0.009869 atm
inch of mercury (at 60°F) = 3.3769 kPa   kPa = 0.296 in. of Hg
inch of water (at 60°F) = 248.6 Pa   Pa = 0.004 in. of water
$lbf/ft^2 = 47.88 \ Pa$ $Pa = 0.02 \ lbf/ft^2$
$lbf/in.^2 = 6.8948 \ kPa$ $kPa = 0.145 \ lbf/in^2$

v. Temperature equivalent

$t_k = (t_f + 459.67)/1.8$ $t_f = 1.8 \ t_k - 459.67$
$t_c = (t_f - 32)/1.8$ $t_f = 1.8 \ t_c + 32$

w. Velocity (length per unit time)

ft/h = 0.085 mm/s   mm/s = 11.76 ft/h
ft/min = 5.08 mm/s   mm/s = 0.197 ft/min
ft/s = 0.3048 m/s   m/s = 3.28 ft/s
in./s = 0.0254 m/s   m/s = 39.37 in./s
mi/h = 0.447 m/s   m/s = 2.24 mi/h

x. Volume

$ft^3 = 0.028 \ m^3 = 28.317 \ L$ $m^3 = 35.71 \ ft^3$
$in.^3 = 16 \ 378 \ mL$ $mL = 0.061 \ in.^3$
gal = 3.785 L   L = 0.264 gal
oz = 29.574 mL   mL = 0.034 oz
pt = 473.18 mL   mL = 0.002 pt
qt = 946.35 mL   mL = 0.001 qt
$acre/ft = 1233.49 \ m^3$ $m^3 = 0.00081 \ acre/ft$

y. Volume per unit time (flow)

$ft^3/min = 0.472 \ L/s$ $L/s = 2.12 \ ft^3/min$
$in.^3/min = 0.273 \ mL/s$ $mL/s = 3.66 \ in.^3/min$
gal/min = 0.063 L/s   L/s = 15.87 gal/min

Temperature Conversion Factors

The numbers in the center column refer to the known temperature, in either °F or °C, to be converted to the other scale. If converting from °F to °C, the number in the center column represents the known temperature, in °F, and its equivalent temperature, in °C, will be found in the left column. If converting from °C to °F, the number in the center represents the known temperature, in °C, and its equivalent temperature, in °F, will be found in the right column.

°C	Known Temp. °F or °C	°F	°C	Known Temp. °F or °C	°F
−59	−74	−101	−42.8	−45	−49.0
−58	−73	−99	−42.2	−44	−47.2
−58	−72	−98	−41.7	−43	−45.4
−57	−71	−96	−41.1	−42	−43.6
−57	−70	−94	−40.6	−41	−41.8
−56	−69	−92	−40.0	−40	−40.0
−56	−68	−90	−39.4	−39	−38.2
−55	−67	−89	−38.9	−38	−36.4
−54	−66	−87	−38.3	−37	−34.6
−54	−65	−85	−37.8	−36	−32.8
−53	−64	−83	−37.2	−35	−31.0
−53	−63	−81	−36.7	−34	−29.2
−52	−62	−80	−36.1	−33	−27.4
−52	−61	−78	−35.5	−32	−25.6
−51	−60	−76	−35.0	−31	−23.8
−51	−59	−74	−34.4	−30	−22.0
−50	−58	−72	−33.9	−29	−20.2
−49	−57	−71	−33.3	−28	−18.4
−49	−56	−69	−32.8	−27	−16.6
−48	−55	−67	−32.2	−26	−14.8
−48	−54	−65	−31.6	−25	−13.0
−47	−53	−63	−31.1	−24	−11.2
−47	−52	−62	−30.5	−23	−9.4
−46	−51	−60	−30.0	−22	−7.6
−45.6	−50	−58.0	−29.4	−21	−5.8
−45.0	−49	−56.2	−28.9	−20	−4.0
−44.4	−48	−54.4	−28.3	−19	−2.2
−43.9	−47	−52.6	−27.7	−18	−0.4
−43.3	−46	−50.8		*(continued next page)*	

°C	Known Temp. °F or °C	°F	°C	Known Temp. °F or °C	°F
−27.2	−17	1.4	−5.6	22	71.6
−26.6	−16	3.2	−5.0	23	73.4
−26.1	−15	5.0	−4.4	24	75.2
−25.5	−14	6.8	−3.9	25	77.0
−25.0	−13	8.6	−3.3	26	78.8
−24.4	−12	10.4	−2.8	27	80.6
−23.8	−11	12.2	−2.2	28	82.4
−23.3	−10	14.0	−1.7	29	84.2
−22.7	−9	15.8	−1.1	30	86.0
−22.2	−8	17.6	−0.6	31	87.8
−21.6	−7	19.4	0	32	89.6
−21.1	−6	21.2	0.6	33	91.4
−20.5	−5	23.0	1.1	34	93.2
−20.0	−4	24.8	1.7	35	95.0
−19.4	−3	26.6	2.2	36	96.8
−18.8	−2	28.4	2.8	37	98.6
−18.3	−1	30.2	3.3	38	100.4
−17.8	0	32.0	3.9	39	102.2
−17.2	1	33.8	4.4	40	104.0
−16.7	2	35.6	5.0	41	105.8
−16.1	3	37.4	5.6	42	107.6
−15.6	4	39.2	6.1	43	109.4
−15.0	5	41.0	6.7	44	111.2
−14.4	6	42.8	7.2	45	113.0
−13.9	7	44.6	7.8	46	114.8
−13.3	8	46.4	8.3	47	116.6
−12.8	9	48.2	8.9	48	118.4
−12.2	10	50.0	9.4	49	120.2
−11.7	11	51.8	10.0	50	122.0
−11.1	12	53.6	10.6	51	123.8
−10.6	13	55.4	11.1	52	125.6
−10.0	14	57.2	11.7	53	127.4
−9.4	15	59.0	12.2	54	129.2
−8.9	16	60.8	12.8	55	131.0
−8.3	17	62.6	13.3	56	132.8
−7.8	18	64.4	13.9	57	134.6
−7.2	19	66.2	14.4	58	136.4
−6.7	20	68.0	15.0	59	138.2
−6.1	21	69.8			*(continued next page)*

°C	Known Temp. °F or °C	°F	°C	Known Temp. °F or °C	°F
15.6	60	140.0	36.1	97	206.6
16.1	61	141.8	36.7	98	208.4
16.7	62	143.6	37.2	99	210.2
17.2	63	145.4	37.8	100	212.0
17.8	64	147.2	38	100	212
18.3	65	149.0	43	110	230
18.9	66	150.8	49	120	248
19.4	67	152.6	54	130	266
20.0	68	154.4	60	140	284
20.6	69	156.2	66	150	302
21.1	70	158.0	71	160	320
21.7	71	159.8	77	170	338
22.2	72	161.6	82	180	356
22.8	73	163.4	88	190	374
23.3	74	165.2	93	200	392
23.9	75	167.0	99	210	410
24.4	76	168.8	100	212	414
25.0	77	170.6	104	220	428
25.6	78	172.4	110	230	446
26.1	79	174.2	116	240	464
26.7	80	176.0	121	250	482
27.2	81	177.8	127	260	500
27.8	82	179.6	132	270	518
28.3	83	181.4	138	280	536
28.9	84	183.2	143	290	554
29.4	85	185.0	149	300	572
30.0	86	186.8	154	310	590
30.6	87	188.6	160	320	608
31.1	88	190.4	166	330	626
31.7	89	192.2	171	340	644
32.2	90	194.0	177	350	662
32.8	91	195.8	182	360	680
33.3	92	197.6	188	370	698
33.9	93	199.4	193	380	716
34.4	94	201.2	199	390	734
35.0	95	203.0	204	400	752
35.6	96	204.8	210	410	770

References

Amerex Corporation, P.O. Box 81, Trussville, AL 35173-0081.

American Society of Plumbing Engineers (ASPE). 1992. Hangers and supports. Chapter 13 in *Data Book.*

Aurora Pump, 800 Airport Road, North Aurora, IL 60542. Fire-protection pump curves.

Coon, J. Walter, P.E. 1991. *Fire protection design criteria option selection.* R.S. Means Company.

Grinnell Corporation. *Sprinkler systems and equipment.* Exeter, NH: Grinnell Corporation, 3 Tyco Park, Exeter, NH 03833.

Heald, C.C., ed. 1988. *Cameron hydraulic data.* Ingersoll-Rand Company.

n.a. 1979. *Private fire protection and detection.* Oklahoma: Oklahoma State University.

National Fire Protection Association (NFPA). 1989. *Carbon dioxide extinguishing systems,* Standard no. 12-1989. Quincy, MA 02269.

———. 1990. *Fire tests of building construction and materials,* Standard no. 251-1990.

———. 1991. *Automatic sprinkler systems handbook.* 5th ed.

———. 1991. *Installation of sprinkler systems,* Standard no. 13-1991.

——. 1992. *Standard types of building construction,* Standard no. 220-1992.
Excerpts from the above-listed materials were reprinted with permission. The reprinted materials are not the complete and official position of the National Fire Protection Association on the referenced subject, which is represented only by the standard(s) in its entirety.

The Viking Corporation, 210 N. Industrial Park Road, Hastings, MI 49058.
Sprinkler pictures and technical details.

Index

3-dimensional Class B fires, 288
3-phase motors, 46
3-way valves, symbols for, 270
25 fire rating, 142
30° elbows, 89, 110
45° elbows
    abbreviations, 122
    equivalent length, 106, 110
    friction loss, 262
    symbols for, 269
    in water distribution system, 89
50 fire rating, 142–143
90° bends, friction loss, 263
90° elbows
    abbreviations, 122
    equivalent length, 106
    friction loss, 97, 101, 262
    in water distribution systems, 89

A

A class combustibles, 165, 284
A class extinguishers, 169–172, 281
AB class extinguishers, 281, 286–287
abbreviations
    hydraulic calculations, 122
    symbols list, 268–273
ABC class extinguishers, 281, 286–287
abort switches
    clean-gas fire-protection system specifications, 234
    high-pressure CO_2 system

        specifications, 223–224
above-ground piping, symbols, 268
absorption in fire-retardant treatment, 21
acceleration of gravity, 94
access panels in water fire-protection specifications, 196
accessibility of pipes, 145
accessories for high-pressure CO_2 specifications, 221
acetone fires, 158
acetylene fires, 158
acre/feet, conversion factors, 303
acres, converting to SI units, 301
activating
    fire-suppression systems, 29, 84
    fires, 15–16
actual tank operating pressure in calculations, 51
adhesion, 91
adjustable clips for branch lines, 141
adjustable swivel loop hangers, 141
advantages (carbon dioxide systems), 156
advisory standards, 10. *See also names of specific standards*
AFFF (aqueous-film-forming foam), 166, 170
air
    oxidizing materials in, 15
    temperature and smoke, 17
air aspirators, 166

air blowers, 166
air compressors
    in sprinkler systems, 67
    water fire-protection system
        specifications, 193–194
air-conditioning systems
    as hazard, 174
    fire-emergency modes, 18
    fire-safety design, 23
air exhaust, smoke and, 17
air-exhaust valves, 68
air leakage
    fires and, 173
    smoke and, 17
air pressure
    air-pressurized barriers, 18
    in hydropneumatic tank pressure
        calculations, 51
    inspecting, 147
    testing and maintenance, 149
air-purged enclosures, 173
air shields on detectors, 35
air supply, smoke and, 17
air vents, symbols for, 271
aircraft hanger fires, 300
airlines, fire drills for, 9–10
alarm check valves
    abbreviations for, 122
    activating alarms, 39
    installation, 209
    sprinkler systems, 74
    symbols for, 270
    water fire-protection system
        specifications, 190, 191
alarm systems
    as part of fire-protection systems,
        6
    carbon dioxide systems, 156–157
    components, 29
    diesel pumps, 47
    fire safety design, 23
    overview, 39
    for sprinkler systems, 68, 75, 85,

86–87
    testing, 148
    water fire-protection system
        specifications, 202
alcohol-resistant foam, 166
alcohol storage area fires, 300
Allendale Insurance, 11
alloys, melting points, 87
alterations to buildings, 27–28
American National Standards
    Institute *ANSI Z210.1*, 301
American Society for Testing and
    Materials (ASTM)
    ASTM E380, 301
    smoke definition, 16
American Water Works Association
    (AWWA), 90
American Welding Society (AWS),
    138
ammunition loading areas or maga-
    zines, fires in, 300
ampere conversion factors, 301
ampere-hours conversion factors, 301
anchors in water fire-protection
    systems, 195
angle valves (AV)
    equivalent length, 106
    friction loss, 261
    in water distribution systems, 89
    water fire-protection system
        specifications, 190
angstroms, 39
annunciators, 29, 75
ANSI Z210.1, 301
antifreeze portable extinguishers, 170
antifreeze sprinkler systems, 68, 139,
    149
applicable publications
    clean-gas fire-protection system
        specifications, 228–229
    high-pressure CO_2 system
        specifications, 217
    portable fire extinguishers

specifications, 213
water system sample specification,
178–179
applications
high-pressure CO_2 system
specifications, 225–226
portable fire extinguishers
specifications, 215–216
water fire-protection system
specifications, 204–205
aqueous-film-forming foam (AFFF),
166, 170
area
conversion factors, 301
of sprinkler system water application, 125
Arkwright Insurance Company, 11
arrangements, sprinkler systems,
143–144
"As-Built Drawings" in water fire-
protection system specifications, 183
asphalt-dipped pipes, 100, 260
asphalt impregnating areas, fires in,
300
aspirators, 166
ASTM (American Society for Testing
and Materials)
ASTM E380, 301
smoke definition, 16
atmospheres (atm), 94, 303
atmospheric pressure, water boiling
points and, 91
audio/visual alarms in high-pressure
CO_2 system specifications, 224
authorities having jurisdiction, 1–3
automatic air vents, 271
automatic dry standpipe system
equipment, 192–193
automatic fire-detection systems, 30
automatic fire-suppression systems,
41
automatic pneumatic detectors, 153

automatic sprinkler systems
installation, 85–86
overview, 65
water-deluge spray systems, 80,
83–85
aviation gas fires, 158
AWS (American Welding Society),
138
AWWA Manual no. 14 (Recom-
mended Practice for Backflow
Prevention and Cross-
Connection Control), 90

B
B class combustibles, 284–285, 288
B class extinguishers, 169–172, 286–
287
backfilling in water fire-protection
system specifications, 205–
206
backflow preventers
common water supplies, 136
diagrams, 296–297
protecting water supply, 90
sprinkler systems, 71, 73–74
symbols for, 270
water fire-protection systems
specifications, 195
ball valves, 261
barriers in buildings, fire safety, 24–
25
base bids in water fire-protection
system specifications, 180
batteries
clean-gas systems, 235
power supplies for high-pressure
CO_2 system specifications,
223
pumps, 47
battery rooms, 300
BC class extinguishers, 281, 286–287
behavioral effects of fire, 7
bells, symbols for, 269

bending movements, conversion
   factors, 301
benzene fires, 158
benzol fires, 158
bibliography, 307–308
bids in water fire-protection system
   specifications, 180
big mouth universal top and bottom
   beam clamps, 141
blankets, 21
bleeders, 155
blocked connections, 147
blocked emergency exits, 147
BOCA (Building Officials and Code
   Administrators International,
   Inc.), 12
boiler room fire doors, 25
boiling points of water, 90, 91
booster-pump systems, 44, 48
borax, 167
Boston
   history of sprinkler systems and,
      66
   Prudential Tower fire, 18
bottled air, 153
brackets
   clean-gas fire-protection system
      specifications, 233
   portable fire extinguishers
      specifications, 215
branches
   arrangement, 143–144
   calculating flow at, 131
   light or ordinary hazard occupan-
      cies and, 113
   number of sprinkler heads per
      branch, 126
   sprinkler system design, 71, 85
brass pipe friction loss tables, 247–
   253
"break glass" stations, 273
breathing apparatus for emergencies
   carbon dioxide system regulations,

   157
high-pressure CO_2 system
   specifications, 225
British thermal units (Btu), 302
Btu (British Thermal units), 302
*Building Construction, Standard
   Types of (NFPA Standard no.
   220)*, 23
Building Officials and Code Administrators International, Inc.
   (BOCA), 12
buildings
   combustibility of construction
      materials, 19
   configuration and smoke, 17
   electrical and mechanical system
      hazards, 174
   fire alarm control panels, 273
   fire barriers, 24–25
   fire safety design, 23–24
   fire-safety personnel, 25–26
   height and smoke, 17
   high-rise fires, 174
   isolating areas, 29
   life safety assessments, 173
   new construction fire safety, 26–27
   remodeling and safety, 27–28
   supports and safety design, 23
buoyancy of smoke, 17
butadiene fires, 158
butane fires, 158
butt-welded elbows, 263
butterfly valves
   abbreviations for, 122
   equivalent length, 106
   friction loss, 262
   symbols for, 270
   in water distribution systems, 89
   water fire-protection system
      specifications, 191
butterfly wafer check valves (WCV),
   122
BV (butterfly valves). *See* butterfly

valves

C

C (carbon), 15, 20
C-clamps, 141
C class extinguishers, 169–172, 286–287
cabinets
  fire hoses and extinguishers, 200–202
  sprinklers, 203
  symbols for, 273
  water fire-protection system specifications, 187
cables
  in detectors, 31–32
  extinguishing fires in, 42
calculations
  booster pump size, 48
  in clean-gas fire-protection system specifications, 228
  equivalent length, 105–107
  fittings and friction loss, 96, 261–266
  gathering information for hydraulic calculations, 121–125
  gravity systems, 108–110
  head/psi tables, 267
  in high-pressure CO_2 system specifications, 219
  hydropneumatic tank pressures, 50–51, 57
  jockey pump starting pressures, 50
  local carbon dioxide application systems, 155
  measurement unit conversions, 301–306
  measurement units in calculations, 94
  in ordinary hazard sprinkler system plans, 126–134
  pipe friction loss tables, 247–260
  pipe-schedule method, 116–117

remote area calculations, 121
sizing carbon dioxide system pipes, 159–163
sizing holding tanks to reservoir pumps, 98–103
straight pipe flow, 96
total head, 46
velocity, 94
velocity head, 92, 97
vertical pipe pressure loss, 98
capacity
  pumps, 45, 46, 57
  spare pumps, 48
capped lines, 269
carbon (C), 15, 20
carbon dioxide (CO_2)
  as pressurizing agent, 167
  extinguishing fires, 21, 42
  portable fire extinguishers, 169, 170
  properties, 151
  in water, 90
Carbon Dioxide Extinguishing Systems (NFPA Standard no. 12), 220
carbon dioxide systems
  advantages and disadvantages, 156
  control panels, 273
  cylinders and scales, 157–159
  fire-suppression systems, 44, 151–154
  flooding factors, 159–160, 162
  high-pressure system specifications, 217–226
  overview, 151
  portable fire extinguishers, 169, 170
  pre-installation information lists, 275–279
  pressure-relief venting formula, 160–161
  sizing pipes, 159–163

specifications, 157
types of systems, 154–156
carbon disulfide fires, 158
carbon monoxide fires, 158
carburetor overhaul shop fires, 300
casings on pumps, 46
cast-iron piping, 100, 260
catalysts for fires, 16
ceiling and wall plate specifications,
    209
ceiling fire ratings, 24
center (eye) of impellers, 46
center load clamps, 141
centigrade conversion factors, 304–
    306
centimeters, 301
centipoise, 91
centistokes, 91
central annunciator panels, 29
centrifugal pumps, 45
CFC (chlorofluorocarbons), 163
chain-operated valves, 271
changes to system specifications,
    184
charts of valves, 203
check valves
    clappers in dry-pipe valves, 84
    in clean-gas fire-protection system
        specifications, 233
    in sprinkler systems, 74
    swing-check valves. *See* swing-
        check valves
    symbols for, 270
    in water distribution systems, 89
    in water fire-protection system
        specifications, 190, 191
chemical extinguishers
    dry chemicals, 167
    extinguishing fires, 21
    halon, 163
    portable fire extinguishers, 169
    wet chemicals, 168
chemistry

chemical properties of water, 90–93
chemical reactions in fires, 15–16
    of fires, 18–20
children, fire education and, 7
chlorofluorocarbons (CFC), 163
circle-marked extinguishers, 171
circuit breakers
    extinguishing fires, 300
    oil-filled, 25
    pumps and, 48
CK.V. *See* check valves
clamps for sprinkler systems, 74
clappers, 84
Class A combustibles, 165, 284
Class A extinguishers, 281, 286–287
Class AB extinguishers, 281, 286–287
Class ABC extinguishers, 281, 286–
    287
Class B combustibles, 284–285, 288
Class B extinguishers, 286–287
Class BC extinguishers, 281, 286–287
Class C extinguishers, 286–287
Class D extinguishers, 286–287
*Clean Agent Fire Extinguishing
    Systems (NFPA Standard no.
    2001),* 163
clean agents, 44, 163, 170
clean-gas fire-protection system
    specifications, 227–237
cleaning fire-protection systems, 148
cleaning plant equipment fires, 300
cleanup
    after fires, 42
    after installation, 185
clearances around pipes, 138
Clevis hangers, 141
close return bends
    equivalent length, 106
    friction loss, 263
closed valves, inspecting, 147
cloth or textiles
    combustibility, 19
    extinguishing fires, 42

history of sprinkler systems and, 66
portable fire extinguishers, 169
clouds (carbon dioxide), 152
CO_2 (carbon dioxide). *See* carbon dioxide (CO_2)
coach screws, 141
coal fires, 158
coaxial cables in detectors, 31
codes and standards
    authorities have jurisdiction, 1–3
    defined, 12
    fire-protection codes and standards, 12–13
    NFPA Standards, 239–246
    state building codes, 12
    state fire-prevention codes and standards, 12
coefficients of discharge or resistance (K), 97, 101, 124, 131, 261
cohesion, 91
cold water
    testing valves, 149
    viscosity and cohesion, 91
color
    hydrant markings, 59
    sprinkler head codes, 77
combination dry-pipe and pre-action sprinkler systems, 68
combination rate-of-rise/fixed-temperature detectors, 32–33
combustible materials
    carbon dioxide systems and, 153
    chemistry of, 15, 18–19
    classification of, 42
    dust, 173
    eliminating, 3
    foam and, 165
    hydrants and, 59
    new construction and, 27
    storage of, 4
combustible metals
    extinguishing fires, 42

portable fire extinguishers, 169
combustion process, 15
combustion products, 36
common water supplies, 136
communications systems
    as hazards, 174
    fire suppression and, 29
compressed air in sprinkler systems, 67
compressibility of water, 90
compressors, air
    in sprinkler systems, 67
    water fire-protection system specifications, 193–194
computer programs (hydraulic calculations), 117, 289–294
computer room fires, 300
computerized smoke-control models, 17
concealed sprinklers, 77, 80
conductance of water, 90
congested rooms, 36
connections
    joints in fire pipes, 138
    standpipe systems, 62
conservation building practices and fires, 173
construction materials, combustibility of, 19
construction methods
    clean-gas fire-protection systems, 237
    high-pressure CO_2 system specifications, 225–226
    portable fire extinguishers specifications, 215–216
    water fire-protection systems specifications, 204–210
construction zone fire safety, 26–27
contaminants, fire and, 16
continuous line detectors, 31–32
contraction of pipes, 89, 95, 110
contractor's sheds, 27

control panels
    clean-gas fire-protection system
        specifications, 234–236
    high-pressure CO$_2$ system
        specifications, 223
    symbols for, 273
control valves
    inspecting, 140
    sprinkler systems, 74
    testing and maintenance, 149
    underground, 271
    in valve boxes, 271
    in water distribution systems, 89
controlling fires, 20–21. *See also*
        extinguishing agents
converting
    measurement unites, 301–306
    temperatures, 304–306
conveyors, as hazards, 174
cooking grease fires, 287
cooling agents, 43, 152
copper piping
    friction loss tables, 247–253
    light and ordinary occupancy pipe
        sizing, 114–115
correction factors in pressure-relief
        venting formula, 160–161
corridors, fire safety design of, 24
corrosion
    corrosion-resistant sprinklers, 80
    protecting against, 137–138
costs
    detectors, 37
    fire damage, 7
    in specifications, 185
coulombs, 301
counting-zone smoke detection, 235–
        236
covered trenches, 162
Cr (crosses), 89, 122
cross mains
    arrangement, 143–144
    sprinkler system design, 71

crosses, 89, 122
cubic feet, 94, 303
cubic inches, 303
cubic meters, 303
cubic milliliters, 303
cushioning, combustibility of, 19
cutting equipment, fire safety and, 26
CV (check valves). *See* check valves
cyclopropane fires, 158
cylinders
    clean-gas fire-protection system
        specifications, 232
    cylinder batteries in high-pressure
        CO$_2$ system specifications,
        157–159, 222
    cylinder brackets in clean-gas fire-
        protection system specifi-
        cations, 233
    reserve cylinders in high-pressure
        CO$_2$ system specifications,
        220
    storage cylinders for clean-gas
        fire-protection system
        specifications, 232
    weight of CO$_2$ cylinders, 159

D

D class extinguishers, 169, 286–287
dampers, closing in carbon dioxide
        systems, 156
Darcy-Weisbach formula, 99–100,
        103, 247–260
deaths, fire-caused, 7
decay period in fires, 173
decreased visibility in carbon dioxide
        systems, 156
"Definitions" section in water system
        sample specification, 179
deflectors on sprinklers, 79, 138
degrees, 303
Del V (deluge valves). *See* deluge
        valves
deluge systems

alarms, 87
  overview, 80, 83–85
  risers and, 121
  sprinkler systems, 67–68
  testing and maintenance, 149
deluge valves
  abbreviations, 122
  overview, 67, 84–85
  sprinkler systems, 74
  symbols for, 271
  water fire-protection system
      specifications, 192
demand at hose stations, 60, 61
density
  density curves, 120
  in hydraulic calculations, 125
  mass per unit volume, 303
  of water, 90, 91, 93
  wet-type sprinkler systems, 120
"Description" section
  clean-gas fire-protection system
      specifications, 227–228
  water system sample specification,
      177–178
design criteria
  designing for fire safety, 23–24
  high-pressure CO_2 system
      specifications, 219–220
  water fire-protection system
      specifications, 180–182
design documents, fire protection
      and, 69
detection systems. *See* detectors; fire
      detection and alarm systems
detectors
  choosing detectors, 36–38
  high-pressure CO_2 system
      specifications, 224–225
  location and spacing, 37
  smoke detectors, 33–35, 38, 39
  symbols for, 272
  types of, 30–36

diagrams of valves, 203
diesel fire-pump rooms, 25
diesel pump motors, 45, 46, 47
diethyl ether fires, 158
dimensions of sprinkler heads, 82
dimethyl fires, 158
directional valves in high-pressure
      CO_2 system specifications, 222
disadvantages (carbon dioxide
      systems), 156
discharge
  alarms in carbon dioxide systems,
      156
  coefficients (K), 101, 124, 131, 261
  delays in carbon dioxide system
      regulations, 157
  hoses in clean-gas fire-protection
      system specifications, 233
  nozzles in clean-gas fire-protection
      system specifications, 233–
      234
  points in carbon dioxide systems,
      163
  rates for sprinkler heads, 79
discharge head, 46, 102
dissolved elements and materials in
      water, 90
distribution piping
  carbon dioxide, 153
  sprinkler system network, 87
doors
  fire safety design, 24–25
  inspecting, 147
  opening into rooms, 7
double drivers in pumps, 46
double-shot carbon dioxide systems,
      162–163
dow therm fires, 158, 300
drafts, smoke and, 17
drain valves, 74
drainage for sprinkler systems, 73–74
drains in water fire-protection
      systems specifications, 209

drawings
  as-built drawings, 183
  for clean-gas fire-protection
        system specifications, 230–
        231
  fire-protection design documents,
        69
  for high-pressure CO_2 system
        specifications, 218–219
  sprinkler rooms, 295, 296
  sprinkler systems plans, 118–119,
        125
  symbols, 268–273
  tagging valves in systems, 203
  valve diagrams in system specifi-
        cations, 183–184
  for water fire-protection system
        specifications, 182–183
drills, fire, 3, 9–10
drinking water supply, protecting, 90
drivers, pumps, 46
drops, symbols for, 268
dry-chemical systems
  extinguishing fires, 42
  portable fire extinguishers, 169,
        170
  types of, 167
dry-compound portable fire extin-
        guishers, 170
dry ice, 152
dry pendent sprinklers, 80
dry-pipe systems, 67
  alarms, 87
  drainage, 73
  maintenance, 149
  testing, 148, 149
dry-pipe valves
  defined, 67
  overview, 84–85
  sprinkler systems, 74
  symbols, 270
  testing and maintenance, 149
  water fire-protection system

specifications, 191
dry sprinkler heads, 80, 139
dry standpipes, 61, 192–193
drying oven fires, 300
ducts
  duct-mounted smoke detectors, 18
  extinguishing fires in, 152
dust collectors, 162
dust covers, 27
dust ignition-proof equipment, 173

E
E (90° elbows). *See* 90° elbows
early suppression fast response
        sprinklers (ESFR), 80, 140
earthquake protection for sprinkler
        systems, 138
eddies in pipes, 95
education, fire-safety, 6–10
EE (45° elbow). *See* 45° elbows
efficiency, pump, 57
elbows
  butt-welded elbows, 263
  equivalent length, 106, 110
  flanged elbows, 263
  friction loss, 97, 262
  long-radius elbows, 106, 262
  long-sweep elbows, 89
  long-turn elbows, 122
  symbols for, 269
  in water distribution system, 89
electric motors
  electric-motor squirrel cage, 46
  pumps, 45, 46
electric "temper" switches, 87
*Electrical Code, National (NFPA
        Standard no. 70),* 11
electrical detectors, 153
electrical equipment
  building systems as hazards, 174
  calculations for carbon dioxide
        systems, 161–163
  carbon dioxide systems and, 152

fires, 42, 288
    hazards, 162
    portable fire extinguishers, 169
electrical-switch-gear rooms, 161–163, 300
electricity conversion factors, 301
electronic equipment. *See* electrical equipment
elevation, pressure calculations and, 110
elevators as hazards, 174
emergency exits, 147
emergency generator room fire doors, 25
emergency lighting, 23
enclosures, testing and maintenance, 149
end-suction pumps, 45
endothermic reactions, 15
energy conservation and fire protection, 173
energy conversion factors, 302
engine test cell fires, 300
engineer jackets (pumps), 47
enlargement of pipes, 89, 95, 110
equations
    carbon dioxide systems, 155
    converting measurement units, 301–306
    Darcy-Weisbach formula, 99–100, 103
    friction loss in pipes tables, 247–260
    Hazen-Williams formula, 103–105, 125–126
    local application systems (carbon dioxide), 155
    number of sprinkler heads, 126
    pressure loss, 103–105
    pressure-relief venting formula, 160–161
    sizing carbon dioxide system pipes, 159–163

velocity head, 92, 97
    water turbulence, 92
equivalent length
    fittings and valves, 105–107, 110
    sprinkler pipes, 129
escalators
    as hazards, 174
    extinguishing fires in, 300
escape routes, 10
ESFR (early suppression fast response sprinklers), 80, 140
ethane fires, 158
ethyl alcohol fires, 158
ethyl ether fires, 158
ethylene dichloride fires, 158
ethylene fires, 158
ethylene oxide fires, 158
European Committee for Standardization, 13
evacuation alarms for carbon dioxide systems, 156
evacuation plans, 3, 153, 156–157
example specifications
    clean-gas fire-protection system specifications, 227–237
    fire-protection notes, 279–280
    high-pressure CO_2 system specifications, 217–226
    portable fire extinguishers, 213–216
    water fire-protection systems, 175–210
excavating in water fire-protection system specifications, 205–206
existing piping, symbols for, 268
exits
    blocked emergency exits, 147
    carbon dioxide system regulations, 157
    fire safety design, 23, 24
    inspecting, 147
    locked, 8

opening into rooms, 7
portable fire extinguishers, 172
exothermic reactions, 15
expansion of pipes, 89, 95, 110
explosion-proof motors, 174
explosions
    combustion reaction, 15
    detecting, 35
explosive fires, 300
exposure fires, 139
extended-coverage sprinklers, 79, 140
extinguishing agents
    carbon dioxide, 152
    cooling agents, 43
    dry chemicals, 167
    foam, 165–167
    halon, 163
    overview, 41–44
    selecting, 299–300
    types of, 42
    wet chemicals, 168
extinguishing fires, methods, 20–21
extra-hazard occupancies
    area per sprinkler, 85
    defined, 5, 6
    fire extinguishers, 283
    pipe sizing, 115
    sprinkler/floor area limitations, 73
extra sprinklers, 139, 211. *See also*
        spare parts
extra-strong steel piping, 254–260
eye nuts, 141
eye rods, 141
eyes of impellers, 46

F
Factory Mutual (FM)
    equipment lists, 2
    Factory Mutual Approval Guide,
        11
    Factory Mutual Data Sheets, 11
    overview, 11–12
Fahrenheit, 304–306

failures in sprinkler systems, 87
false alarms, 31, 37, 38
farads, 301
fast-growing fires, 66
fast-response sprinklers, 80, 140
fc (footcandles), 302
FD (fire-department connections).
        See fire-department connec-
        tions
federal authorities, 1
feed mains in sprinkler systems, 71
fees in water fire-protection system
        specifications, 186
feet
    converting to SI units, 301, 302
    feet per second (fps), 92
female connections, standpipes, 62
fighting fires. *See* fire protection and
        suppression systems
finishes, combustibility of, 19
fire alarm systems. *See* alarm systems
fire barriers, 24–25
fire codes, defined, 12
fire dampers, 18
fire-department connections
    blocked, 147
    installation in specifications, 209
    overview, 135–136
    in pipe-schedule method, 117
    in specifications, 197, 202
    sprinkler systems, 75
    symbols for, 271, 272
    testing and maintenance, 149
fire-department inlets, 271
fire department responses per day, 7
fire detection and alarm systems
    alarms, 39
    as part of fire-protection systems,
        6
    automatic pneumatic detectors,
        153
    choosing detectors, 36–38
    components, 29

flame detectors, 35–36, 38, 39
gas detectors, 39
HAD (heat-actuator detectors),
    153
inspecting detectors, 145
manual and automatic, 30
overview, 29–30
smoke detectors, 33–35, 38, 39
symbols for, 272
testing and maintenance, 149
types of detectors, 30–36
fire development (speed), 36
fire drills, 3, 9–10
fire-emergency modes (air condition-
    ing), 18
fire extinguishers
    guidebooks, 281–294
    halon, 163
    maintenance, 148
    new construction, 27
    portable fire extinguishers, 169–
        172
    selecting, 282–285, 299–300
    symbols for, 272
    in water fire-protection system
        specifications, 201–202
*Fire Extinguishers, Portable (NFPA
    Standard no. 10)*, 171, 172
fire hose cabinets. *See* hose cabinets
    and racks
fire hoses
    fire hose racks, 272
    fire hose valves, 270
    in water fire-protection system
        specifications, 201–202
fire hydrants. *See* hydrants
fire insurance coverage, 11
fire load, 143
fire-loss prevention-and-control
    manager, 26
fire marshals, 2
fire meters, 272
fire prevention

as part of fire-protection systems,
    6
energy conservation building
    practices and, 173
organizations, 10–12
fire protection and suppression
    systems
    activating, 29
    in buildings, 25–26
    carbon dioxide. *See* carbon dioxide
        systems
    clean-gas fire-protection system
        specifications, 227–237
    codes and standards, 12–13
    defined, 4
    definitions, 142
    design documents, 69
    detection and alarm systems. *See*
        fire detection and alarm
        systems
    dry-chemical systems, 167
    extinguishing agents, 41–44
    general precautions, 3–6
    halon, 163
    high-pressure CO_2 system
        specifications, 217–226
    inspection schedules, 145–146
    maintenance, 146–149
    organizations, 10–12
    overview, 41
    piping loops, 59
    portable fire extinguishers
        specifications, 213–216
    prevention, 6, 10–12, 173
    projects, 2–3
    smoke control, 17–18
    sprinkler systems. *See* sprinkler
        systems
    suppression, 6
    types of carbon dioxide systems,
        154–156
    water-suppression systems, 59–63
    water system sample specification,

175–210
wet-chemical systems, 168
fire-protection specialists
  chemistry knowledge, 18
  defined, 11
  fire-prevention officials, 2
fire pumps
  booster pumps, 48
  capacities, 45
  components, 46–48
  dimensions, 56
  fire-pump rooms, 57
  hydropneumatic tanks, 50–51, 57
  jockey pumps, 49–50
  maintaining pressure, 49–57
  overview, 45–46
  pump curves, 52–58
  spare pumps, 48–49
  in specifications, 182, 198–200
  symbols for, 271
fire ratings
  25 and 50 ratings, 142–143
  building materials, 24
  fire-rated doors, 24–25
  NFPA fire-resistance ratings, 23
fire resistancy
  fire-resistant coatings, 21
  fire-resistant floors, 24
  fire-retardant materials, 21
  NFPA fire-resistance ratings, 23
fire safety
  building design, 23–24
  education, 6–10
  NFPA efforts, 10–11
  personnel, 25–26
  plans, 3, 147
fire signatures, 16, 173
fire suppression. *See* fire protection
  and suppression systems
fire triangle, 15–16, 42
fire walls, 24
fires
  chemistry, 18–20

development of, 173
fire triangle, 15–16, 42
higher temperatures, 173
methods of extinguishing, 20–21
number per day, 7
selecting extinguishing agents,
  299–300
signature, 16, 173
smoke, 16–18
statistics, 10
types of, 281
uncontrolled, 1
firetrap buildings, 7
first detection alarms in high-
  pressure CO_2 system specifications, 224
fittings
  defined, 89
  equivalent length, 105–107
  flow calculations, 96
  friction loss, 261–266
  head loss, 99, 101, 102
  high-pressure CO_2 system
    specifications, 221
  sprinkler systems, 74
  in water fire-protection system
    specifications, 190
fixed foam makers, 166
fixed monitors (foam systems), 166
fixed nozzles on hydrants, 59
fixed-temperature detectors, 31–32,
  38, 139
fixed water systems, 43
flame detectors, 30, 35–36, 38, 39
flame-resistant. *See* fire resistancy
flame spread ratings, 142
flammable circuit breakers, 25
flammable gases, 43
flammable liquids
  carbon dioxide systems and, 152
  extinguishing fires, 42
  extinguishing fires in storage
    areas, 300

foam, 165
  portable fire extinguishers, 169,
    288
flammable solids storage, 300
flanged elbows, 263
flanged lines, 269
flash points, 16
flooding
  factors for specific hazards, 162
  types of carbon dioxide systems,
    155
floors
  areas protected by sprinklers, 72–
    73
  fire safety design, 24
  isolating, 174
  rubber coating, 174
flow
  clean-gas fire-protection system
    calculations, 228
  fittings and, 96
  hydrant classifications, 59
  hydraulic calculations, 117–134
  increments (q), 122
  mass per unit time, 302
  sprinkler system rates, 73
  volume per unit time, 303
  water-deluge systems, 83
  water fire-suppression systems, 44
  water in pipes, 92–93
flow switches, 273
fluid dynamics, 93
fluoroprotein compounds in foam,
    165, 166
flush mounted fixtures, 268
flush pipe entrances, 266
flush sprinklers, 77, 80
flushing pipes
  flushing rates, 72
  testing and maintenance, 149
  underground pipes, 208
  water fire-protection systems
    specifications, 209

FM (Factory Mutual Research
    Corporation), 2, 11–12
foam
  combustibility, 19
  concentrate, 165
  extinguishers, 165–167
  extinguishing fires, 42
  portable extinguishers, 170
fog streams (water), 63
fogs (carbon dioxide), 152
food-processing areas and kitchens,
    152, 168
foot valves, 101, 265
footcandles (fc), 302
force conversion factors, 302
force per unit length, 302
formulas. *See* equations
fps (feet per second), 92
frangible bulb sprinkler components,
    75–76, 77
frangible pellet sprinkler components,
    75–76
freezing temperatures
  dry-pipe systems and, 67
  freeze-ups, 147
  freezing points of water, 90
  man-made reservoirs and, 43
  protecting pipes against, 137
  sprinkler systems and, 66
freight and hauling in water fire-
    protection system specifica-
    tions, 184–185
fresh-air dampers or doors, 29
friction as fire activation, 15
friction losses in flow
  calculating, 94, 122
  Darcy's formula, 99–100
  discharge side of pumps, 102
  in fittings, 261–266
  Hazen-Williams formula, 103–105,
    125–126
  in pipes, 92, 247–260
  pressure loss defined, 89, 94

vertical pipes, 98
viscosity and, 91
in water systems, 99, 101
FS (flow switches), 273
ft³ (cubic feet), 94, 303
fuel oil
    fuel-oil pump rooms, 25
    pumps, 47
    storage fires, 300
fuels
    cutting off supplies in gas fires, 43
    molecules, 15
    removing from fires, 20
fully-developed period in fires, 173
fur-storage vaults, 162
fusible-link sprinkler components, 75,
    76–77

G

gallons (gal), 94, 303
gallons per minute (gpm), 46, 52–58,
    122
galvanized piping for sprinkler
    systems, 70
gas-fire suppression systems
    clean-gas fire-protection system
        specifications, 227–237
    inspecting, 145
gas fires, 166
gases
    carbon dioxide systems and, 152,
        156
    combustibility, 19
    extinguishing fires, 43
    gas detectors, 39
gasoline fires, 158
gate valves (GV)
    abbreviations, 122
    equivalent length, 106, 110
    friction loss, 97, 261
    symbols for, 270
    in water-distribution systems, 89
    in water fire-protection system

specifications, 190
gauges
    inspecting, 147
    symbols for, 272
general design criteria in water fire-
        protection system specifica-
        tions, 180–182
general materials sections
    in portable fire extinguishers
        specifications, 214
    in water fire-protection system
        specifications, 186–187
general requirements
    in clean-gas fire-protection system
        specifications, 229–230
    in high-pressure CO_2 system
        specifications, 217–218
    in portable fire extinguishers
        specifications, 213–214
general sections
    in clean-gas fire-protection system
        specifications, 227–231
    in water system sample specifica-
        tion, 177–186
*General Storage (NFPA Standard
    no. 231)*, 69
geometry of rooms, 36
globe valves
    equivalent length, 106
    friction loss, 261
    symbols for, 270
    in water-distribution systems, 89
    in water fire-protection system
        specifications, 190
gongs, symbols for, 269
gpm (gallons per minute), 46, 52–58,
    122
gravity
    acceleration of water, 94
    calculating pressure, 108–110
    in water fire-suppression systems,
        44
grease fires, 42, 287

gridded arrangement (sprinkler systems), 143–144

grounding, in fire safety design, 23

growth period in fires, 173

guarantees in water fire-protection system specifications, 186

guards and shields on sprinkler pipes, 137

guides for pipes, 195

GV (gate valves). *See* gate valves (GV)

H

H (hydrogen), 15, 20, 158

HAD (heat-actuator detectors), 153

hallways, fire safety design of, 24

halon

    fire extinguishers, 163

    fire-suppression systems, 44

    halon 1211, 172

    halon 1301, 172

    halon replacement extinguishers, 21

hand-held lines

    carbon dioxide systems, 155

    dry chemical systems, 167

    foam systems, 166

handling

    in clean-gas fire-protection system specifications, 231

    in high-pressure CO_2 system specifications, 221

    in portable fire extinguishers specifications, 214

hangar decks, extinguishing fires on, 300

hangers, 74, 141, 142, 194–195

hardness of water, 90

hauling in water fire-protection system specifications, 184–185

hazardous materials, 3

hazards

    flooding factors for specific hazards, 162

    NFPA hazard classification, 4–6

    selecting extinguishing agents, 299–300

Hazen-Williams formula, 103–105, 125–126

head. *See* pressure

head loss. *See* pressure drops or differences

heat

    as fire activation, 16

    consumption in fires, 15

    defined, 15

    detecting rapid rises, 32–33

    extinguishing effects of carbon dioxide, 152

    structural failure and, 24

    triggering sprinkler systems, 68

    water absorption, 20

heat-actuator detectors (HAD), 153

heat detectors

    outdoor protection, 139

    symbols for, 271, 273

    types of detectors, 30, 31–33, 38, 39, 153

heat-sensitive semiconductor insulation, 31

heated pump rooms, 137

heaters

    heater room fire doors, 25

    heating systems as hazards, 174

    in new construction, 27

hectares, 301

henrys, 301

hexane fires, 158

high-expansion foam, 166

high-hazard areas, 67

high-piled storage, 72

high-pressure CO_2 systems

    high-pressure containers, 153, 154, 157–159

    sample specification, 217–226

high-rise buildings
  central annunciator panels, 29
  firefighting in, 174
high-temperature sprinkler heads
  symbols for, 269
  testing and maintenance, 140, 149
high-velocity fire plumes, 139–140
high-velocity spray nozzles, 83
higher paraffin hydrocarbon fires, 158
history
  portable fire extinguishers, 172
  sprinkler systems, 66
holding tank pump sizing calcula-
    tions, 98–103
horizontal pumps, 45
horizontal sidewall sprinklers, 79, 81
horizontal split-case pumps, 45
horns
  high-pressure CO_2 system
      specifications, 224
  symbols for, 269
horsepower, 57, 303
hose cabinets and racks
  fire hose racks, 272
  fire hoses and extinguishers, 200–
      202
  in specifications, 187, 200–201
  symbols for, 272, 273
hose-station systems
  as fixed water systems, 43
  classifications, 60, 61
  distances between, 63
  stagnant water in, 90
  water supply pressure, 70
hose stream demand, 60, 61
hose threads in water fire-protection
    system specifications, 188–
    189
hoses
  in carbon dioxide systems, 155
  in clean-gas fire-protection system
      specifications, 233
  defined, 63

fire hose valves, 270
hose cabinets and racks, 187, 200–
    202, 272, 273
in water fire-protection system
    specifications, 201–202
hospital fire drills, 9
host conversion factors, 302
hot water viscosity and cohesion, 91
housekeeping practices, 8
housing on pumps, 46
hp (horsepower), 57, 303
hydrants
  fire-protection piping loop, 59
  fixed water systems and, 43
  hydrant tests, 59, 124
  inspecting, 147
  new construction and, 27
  symbols for, 271
*Hydrants Testing and Marking
    (NFPA Standard no. 291),* 59
hydraulic calculations, 117–134
  defined, 113
  gathering information, 121–125
  samples, 289–294
  software, 117
  in specifications, 180
  worksheets, 133
hydraulic oil, extinguishing fires in,
    300
hydraulics, 89, 93–110
hydro-turbine generator fires, 300
hydrocarbon combustibles, 165
hydrogen (H), 15, 20, 158
hydrogen-ion concentration of water,
    90
hydrogen sulfide, 158
hydropneumatic tanks, 50–51, 57
hydrostatic tests, 135

I

identification in water fire-protection
    system specifications, 195–
    196

illuminance conversion factors, 302
impact protection, 137
impellers
    pumps, 46
    rpm, 57
in-line pumps, 45
in-rack sprinkler demand, 125
in^3 (cubic inches), 303
inches
    converting to SI units, 301, 302
    of mercury, 303
    of water, 303
increasers or reducers, 269
indoor fire detectors, 38
indoor transformers, extinguishing
        fires in, 300
induction type motors, 46
industrial construction, fire doors
        and, 25
Inergen, 42, 44, 169, 172
inert atmospheres, 21
inert gas, 43
inerting rooms, 43
information lists and sheets
    carbon dioxide pre-installation
        information lists, 275–279
    hydraulic calculations, 122, 123
infrared detectors, 35, 38, 39, 272
inhibitors (fire), 16
initial alarms for carbon dioxide
        systems, 156
inlet head, 46
inside-hose demand, 125
inspecting
    control valves, 140
    fire-protection projects, 3, 145–146
    high-pressure CO$_2$ system
        specifications, 226
    inspection reports and records,
        147
    inspections as maintenance, 146–
        147
    inspector's tests in water fire-

protection systems
    specifications, 210
insurance inspection departments,
        1
*Inspection, Testing, and Mainte-
    nance of Sprinkler Systems
    (NFPA Standard no. 13A)*, 65,
    135, 184
inspector's test connections, 269
*Installation of Centrifugal Fire
    Pumps (NFPA Standard no.
    20)*, 58, 179
*Installation of Sprinkler Systems
    (NFPA Standard no. 13). See
    NFPA Standard no. 13*
installing
    carbon dioxide pre-installation
        information lists, 275–279
    clean-gas fire-protection systems,
        237
    high-pressure CO$_2$ system
        specifications, 225–226
    inspection and, 145
    portable fire extinguishers
        specifications, 215–216
    sprinkler systems, 85–86
    water fire-protection system
        specifications, 204–205
insulation
    higher temperatures of fires and,
        173
    insulated cable fires, 42
insurance companies
    fire insurance coverage, 11
    insurance inspection departments,
        1
    requirements, 12
    sprinkler system reliance, 69
interference to sprinkler systems, 85
investigations, NFPA, 10
ionization-type smoke detectors, 33,
        34, 38, 272
isobutane fires, 158

isobutylene fires, 158
isobutylene formate fires, 158
isolating building areas, 29, 174

J
J-hangers, 141
J (joules), 302
jet engine test cell fires, 300
jockey pumps, 49–50
joints in fire pipes, 138
joules, 302
JP-4 fires, 158
jurisdiction, 1–3

K
K (resistance coefficient), 97, 101, 261
K (water-discharge coefficient), 124, 131
kerosene fires, 158
kgf (kilogram-force), 302
kilogram-force (kgf), 302
kilometers (km), 301
kilopascals (kPa), 303
kilowatt hours (kWh), 302
kitchens, 152, 168
kPa (kilopascals), 303
kWh (kilowatt hours), 302

L
labeling
    FM Approval label, 11
    organizations, 2
laminar flow in pipes, 92, 93
large-drop sprinklers, 139–140
large flange clamps, 141
last event recall (clean-gas systems), 235
lb (pounds), 94
lbf (pound-force), 302
lbm (pound-mass), 302
leakage
    compensating for in CO_2 systems, 160

damage in water fire-protection
        system specifications, 184
    during tests, 135
    inspections of leaking pipes, 147
left-hand rotation pumps, 56
length
    conversion factors, 302
    equivalent length, 105–107, 110
    hoses, 63
    in straight pipe flow calculations, 96
life safety assessments, 173
Life Safety Code, 7–9
lift check valves, 89, 265
light
    conversion factors, 302
    wavelengths, 35
light hazard occupancies
    area per sprinkler, 72, 85
    fire extinguishers, 283
    pipe sizing, 114
    sprinkler branch lines, 113
    types of, 4, 6
light-obscuration smoke detectors, 33, 34
light-scattering smoke detectors, 33, 34–35
lighting
    as hazards, 174
    fire safety design, 23
    high-pressure CO_2 system
        specifications, 224
lignite storage area fires, 300
linear acceleration conversion
        factors, 301
lines to be removed, symbols for, 268
liquefied carbon dioxide, 151
liquefied petroleum gas fires, 300
liquids
    combustibility, 19
    organic liquids, 19
listing agencies, 2
listings for pumps, 45, 58

loading docks, 139, 167
local alarms, 29
local application systems (carbon dioxide), 155, 277–279
local authorities, 1
location of detectors, 37
locked exits, 8
locked valves, 140, 149
long-radius elbows
    equivalent length, 106
    friction loss, 262
    long-sweep elbows, 89
    long-turn elbows, 122
long runs of pipe, 137
long sweep elbows, 89
long-turn elbows, 122
looped arrangement (sprinkler systems), 143–144
low-point drains, 149
low-pressure carbon dioxide containers, 153, 154
low-pressure water sources, 48
low-velocity nozzles, 83
Lt. E (long-turn elbows), 122
lubricating hydrants, 60
lubricating oil, 158, 300

M

magnesium
    carbon dioxide systems and, 153
    portable fire extinguishers, 169
magnetism, conversion factors, 301
main drains, 149
mains
    branch arrangements, 143–144
    flushing, 72
    long runs of pipe, 137
    standpipes and, 60
maintenance
    cleaning systems, 148
    fire protection systems, 146–149
    hydrants, 60
    inspections and, 146–147

portable fire extinguishers, 171
preventative maintenance, 148
repairs and replacements, 148
sprinkler systems, 69
testing systems, 148
male connections in standpipes, 62
malleable swivel hangers, 141
manual controls
    manual discharge controls in high-pressure CO_2 system specifications, 220
    on pumps, 48
manual fire alarm pull stations, 222–223, 272
manual fire-detection systems, 30
manual hydraulic calculations, 121
manual water-suppression systems, 59–63
manufacturers data in water fire-protection system specifications, 183
markings
    identification in water fire-protection system specifications, 195–196
    portable fire extinguishers, 171
mass
    conversion factors, 302
    per unit area, 302
    per unit time (flow), 302
    per unit volume (density), 303
match lines, 269
materials
    in clean-gas fire-protection system specifications, 231–236
    in high-pressure CO_2 system specifications, 221–225
    in portable fire extinguishers specifications, 214–215
    in water fire-protection system specifications, 186–204
measurement units
    in calculations, 94

converting, 301–306
mechanical detectors, 153
mechanical systems in buildings, 174
mechanical water flow alarms, 86
medium-expansion foam, 166
melting points, 75, 87
meters, 272
methane fires, 158
methyl acetate fires, 158
methyl alcohol fires, 158
methyl ethyl ketone fires, 158
methyl formate fires, 158
miles, converting to SI units, 301, 302
milliliters, 303
minutes, 303
missing nozzles, 147
mitre bends
    equivalent length, 106
    friction loss, 263
mobile carbon dioxide systems, 155–
    156
moments of inertia, 303
monoammonium phosphate, 167
motor-operated valves, 270
motors
    explosion-proof, 174
    pumps, 45, 46
movement of smoke, 17
mrad, 303
multistage pumps, 45
municipal fire alarm stations, 272

N

N m (Newton-meters), 301
National Electrical Code (NEC), 11
National Fire Protection Association,
    Inc. *See* NFPA (National Fire
    Protection Association, Inc.)
natural gas fires, 158
new construction fire safety, 26–27
new piping
    friction loss tables, 254–260
    symbols, 268

New York Triangle Shirtwaist fire, 7–9
Newton meters, 301
NFPA (National Fire Protection
    Association, Inc.), 1
    address, 13
    list of standards, 239–246
    noncombustible materials, 19
    overview, 10–11
    pipe and tube materials, 143
    rates of occupancy, 4–6
    role of, 2
    types of hazard, 4–6
*NFPA Standard no. 10 (Fire
    Extinguishers, Portable),* 171,
    172
*NFPA Standard no. 12 (Carbon
    Dioxide Extinguishing
    Systems),* 220
*NFPA Standard no. 13 (Installation
    of Sprinkler Systems)*
    color coding, 77
    flushing rates, 72
    history of standard, 65
    hose stations and standpipes, 60,
        61
    maintenance defined, 148
    pipe-schedule method and, 113
    requirements, 135
    in sample specifications, 179
    water density curves, 120
    water-flow alarms, 39
    working design drawing require-
        ments, 69
*NFPA Standard no. 13A (Sprinkler
    Systems, Care, Maintenance),*
    65, 135, 184
*NFPA Standard no. 13D (Sprinkler
    Systems, One- and Two-
    Family Dwellings),* 66
*NFPA Standard no. 13R (Sprinkler
    Systems, Residential Occu-
    pancies up to and including
    Four Stories in Height),* 66

NFPA Standard no. 14 (Standpipe and Hose Systems), 60, 179

NFPA Standard no. 20 (Installation of Centrifugal Fire Pumps), 58, 179

NFPA Standard no. 22 (Water Tanks for Private Fire Protection), 43

NFPA Standard no. 25 (Water-Based Fire Protection Systems), 184

NFPA Standard no. 70 (Electrical Code, National), 11

NFPA Standard no. 72 (Protective Signaling Systems), 37

NFPA Standard no. 92A (Recommended Practice for Smoke Control Systems), 16

NFPA Standard no. 220 (Building Construction, Standard Types of), 23

NFPA Standard no. 231 (Storage, General), 69

NFPA Standard no. 231C (Rack Storage of Materials), 69

NFPA Standard no. 251 (Standard Methods of Fire Tests of Building Construction and Materials), 23

NFPA Standard no. 291 (Hydrants Testing and Marking), 59

NFPA Standard no. 2001 (Clean Agent Fire Extinguishing Systems), 163

nipple assemblies in clean-gas fire-protection system specifications, 234

nitrogen
    inspecting pressure, 147
    in pipes, 67
    pressurizing agent, 167

nomograms (water density), 73

non-combustible materials, 19, 24

non-fire-rated doors, 25

non-potable firefighting water supply, 89

normal pressure (P_n), 122

notes (sample fire-protection notes), 279–280

nozzles
    in clean-gas fire-protection system specifications, 233–234
    in high-pressure CO_2 system specifications, 222
    hoses, 63
    hydrants, 59
    missing, 147
    sprinkler head types, 80
    water-deluge spray systems, 80, 83

numbers
    on extinguishers, 169
    on valves, 140

O

O_2 (oxygen). *See* oxygen

occupancy classification (NFPA), 4–6

Occupational Safety and Health Administration (OSHA), 157

ohms, 301

oil
    extinguishing fires in, 42
    pressure in pumps, 47

oil-filled circuit breakers or switches, 25

oil-quenching bath fires, 300

oil-storage room fire doors, 25

old style sprinklers, 78, 80

on-off sprinkler heads, 80

open flames, 4, 26

open head nozzles, 80

open sprinklers, 80, 149

open valves, 147

openings in buildings, fire safety design, 24

operating instructions
    clean-gas fire-protection system specifications, 230

high-pressure CO_2 system
    specifications, 219
operating pressure in
    hydropneumatic tank pressure
    calculations, 51
operational activities in buildings,
    detectors and, 36
ordinary hazard occupancies
    area per sprinkler, 72, 85
    calculating pipe sizes, 116–117
    defined, 5, 6
    fire extinguishers, 283
    pipe sizing, 114–115
    sprinkler branch lines, 113
    sprinkler system plans, 126–134
organic liquids, 19
organic solids, 20
organizations, fire-protection, 10–12
orifices
    in clean-gas fire-protection system
    specifications, 234
    flow in pipes and, 95
    size on sprinklers, 79, 83, 124
ornamental sprinklers, 80
OSHA (Occupational Safety and
    Health Administration), 157
ounce-force (ozf), 302
ounce-mass (ozm), 302
ounces (oz), 303
outdoor fire detectors, 38
outdoor sprinkler heads, 139
outdoor transformer fires, 300
outdoor water motor gongs, 87
over-pressure vent openings, 160–
    161
overhead piping
    installation, 207–208
    in specifications, 189–190
    testing, 208
overtime work in water fire-protection
    system specifications, 185
owner inspections, 145
oxidizing materials in air, 15, 153

oxygen
    as oxidizing material, 15
    carbon dioxide and, 152
    deficiency with carbon dioxide
    system use, 156
    lowering concentration, 42
    removing from fires, 21
oz (ounces), 303
ozf (ounce-force), 302
ozm (ounce-mass), 302
ozone layer, 163

P
paint fires, 300
paint spray booth fires, 300
painting sprinkler systems, 138
panic, 7
paper
    extinguishing fires in, 42
    portable fire extinguishers, 169
    storage, 162
parking garages, 156
passive restraints, fire safety and, 24
P_e (pressure/elevation), 122
pendent sprinklers, 78, 79, 81, 268
penetrations, fire safety design, 24
pentane fires, 158
permits, 2, 186
personnel, fire-safety, 25–26
petrochemical storage area fires, 300
petrochemical testing laboratory
    fires, 300
P_f (pressure loss due to friction), 122
pH of water, 90
photoelectric smoke detectors, 33,
    34–35, 38, 272
physical fire barriers, 18
physics, 93
pints, converting to SI units, 303
pipe guides in water fire-protection
    systems, 195
pipe hangers, 74, 141, 142, 194–195
pipe-schedule method

calculations, 116–117
defined, 113
pipes and piping. *See also* sizing
  carbon dioxide, 153
    in high-pressure CO_2 system
      specifications, 221
    sizing, 159–163
  clamps, 141
  clean-gas fire-protection system
    specifications, 231–232
  clearances around, 138
  extra hazard occupancy pipe
    sizing, 115
  friction loss
    entrance friction loss, 266
    exit friction loss, 266
    pipe bend friction loss, 263
    tables, 247–260
  hangers, 141, 142, 194–195
  inspecting, 145
  installation specifications, 206–207
  long runs, 137
  materials, 143, 221
  piping loops, 59
  protecting, 137–138
  sizing
    calculations, 94
    carbon dioxide system pipes,
      159–163
    light and ordinary occupancy
      sizing, 114–115
    sprinkler systems, 68, 73, 113–
      116
    standpipe systems, 61–63
  sprinkler systems, 68, 70–71, 73,
    74, 113–116
  standpipe systems, 61–63
  symbols, 268
  water-filled, 142
  in water fire-protection system
    specifications, 189–190,
    206–207
pitched roofs, 134

plane angles, 303
plant air, 153
plastic material combustibility, 19
plated sprinkler heads, 77
plug valves, 262
plumbing as fire hazards, 174
plumbing drawings. *See* drawings
P_n (normal pressure), 122
polar-solvent AFFF concentrate, 166,
  288
portable fire extinguishers
  classifications, 169–172
  dry chemicals, 167
  guidebooks, 281–294
  maintenance, 148, 171
  mobile carbon dioxide systems,
    155–156
  new construction and, 27
  specifications, 213–216
*Portable Fire Extinguishers (NFPA
  Standard no. 10),* 171, 172
portable monitors (foam), 166
positive suction pressure, 46
post-indicator valves, 59, 271
potable water supply, protecting, 90
potassium, carbon dioxide systems
  and, 153
potassium bicarbonate, 21, 167
potassium chloride, 167
pound-force (lbf), 302
pound-mass (lbm), 302
pounds (lb), 94
pounds per square inch (psi), 94, 122,
  124, 267
powder mixtures, 167
power
  as hazards, 174
  conversion factors, 303
pre-action systems
  alarms, 87
  control panels, 273
  defined, 67
  high-pressure CO_2 system

specifications, 220
testing and maintenance, 149
pre-action valves
    symbols for, 270
    water fire-protection system
        specifications, 192
pre-installation information lists, 275–
    279
pressure. *See also* pressure drops or
        differences
    conversion factors, 303
    corresponding psi table, 267
    elevation difference symbols, 122
    fittings and flow calculations, 96
    gravity systems, 108–110
    hydropneumatic tanks, 50–51, 57
    inlet head, 46
    pumps, 46, 49–57
    rated head, 46
    residual pressure, 28
    sprinkler systems water supply, 70,
        71–72
    static head. *See* static head
    straight pipe flow calculations, 96
    suction head, 46
    triple-point pressure (CO_2), 151
    velocity head, 89, 92, 97
    vertical pipe calculations, 98
    water-deluge systems, 83
    water fire-suppression systems, 44
    water pressure, 93
pressure-activated alarm switches,
        86–87
pressure drops or differences
    defined, 89
    friction and, 94
    Hazen-Williams formula, 103–105,
        125–126
    hydraulic calculations, 117–134
    P_f, 122
    pumps, 48, 101, 102
    straight pipe flow, 96
    total pressure drop, 132, 133–134

velocity head, 89, 92, 97
pressure foam makers, 166
pressure gauges
    testing and maintenance, 149
    water fire-protection system
        specifications, 196
pressure impregnation, 21
pressure indicators, 273
pressure loss. *See* pressure drops or
        differences
pressure-regulating or reducing
        valves, 198
pressure-relief venting formula, 160–
        161
pressure switches (PS)
    jockey pumps, 49
    symbols for, 273
pressurizing stairwells, 17, 18
preventative maintenance, 148
priming
    inspecting, 147
    pumps, 47
    water levels, 149
printing press fires, 300
private water systems, 59
product delivery, storage and
        handling
    clean-gas fire-protection system
        specifications, 231
    high-pressure CO_2 system
        specifications, 221
    portable fire extinguishers
        specifications, 214
project development process, 2–3
propane fires, 158
properties of water, 90–93
property damage caused by fire, 7
propylene fires, 158
protecting sprinkler pipes, 137–138
protecting water supply, 90
Protection Mutual Insurance
        Company, 11
protective cages, 268

Protective Signaling Systems (NFPA Standard no. 72), 37
protein compounds in foam, 165
Prudential Tower fire, 18
PRV (pressure-regulating or reducing valves), 198
psi (pounds per square inch), 94, 122, 124, 267
psychological effects of fire, 7
pt (pints), 303
P_t (total pressure), 122, 130
puff tests, 157
pull cables on carbon dioxide containers, 153
pull stations, 272
pump housing, 46
pump rooms, 25, 57
pump test headers, 269
pumps
    booster pumps, 48
    capacities, 45
    components, 46–48
    defined, 45
    dimensions, 56
    fire-pump rooms, 57
    hydropneumatic tanks, 50–51, 57
    jockey pumps, 49–50
    maintaining pressure, 49–57
    overview, 45–46
    pump curves, 52–58
    pump-head values, 107–108
    sizing, 48, 98–103, 107–108
    spare pumps, 48–49
    in specifications, 182, 198–200
    symbols for, 271
    water fire-suppression systems, 44, 198–200
punch lists, 145
purple K, 167
P_v (velocity pressure), 122

Q
Q (pump capacity), 45, 46, 57
Q (summation of flow), 122, 129
QREC (quick-response extended coverage sprinklers), 80
QRES (quick-response early suppression sprinklers), 80, 149
qt (quarts), 303
quality assurance
    high-pressure CO_2 system specifications, 221
    water fire-protection system specifications, 187
quarts (qt), 303
quench fires, 158
quick-response sprinklers
    illustrated, 83
    quick-response early suppression sprinklers (QRES), 80, 149
    quick-response extended coverage sprinklers (QREC), 80
    types of sprinklers, 80

R
Rack Storage of Materials (NFPA Standard no. 231C), 69
µrad, 303
radiant energy, 35
rapid rises in temperature, 32–33
rate-anticipation detectors, 139
rate-compensation detectors, 31, 32, 33, 38
rate-of-rise detectors, 31, 32–33, 38, 139
rated head (pumps), 46
rating bureaus, 1
ratings
    fire ratings, 142–143
    portable extinguishers, 169
reactive metals, carbon dioxide systems and, 153
reactor and fractionating tower fires, 300

recessed sprinklers, 77, 80
Recommended Practice for Backflow
Prevention and Cross-
Connection Control (AWWA
Manual no. 14), 90
Recommended Practice for Smoke
Control Systems (NFPA
Standard no. 92A), 16
records
bulk paper, 162
of fire drills, 3
of inspections, 147
record vault fires, 300
reducers, 269
references, 307–308
refrigerated area drainage, 73
refuse disposal, as hazard, 174
regional authorities, 1
reinspecting sprinkler systems, 28
related work in water system sample
specifications, 178–179
reliability
fire-protection systems, 146
of water systems, 43
relief vent fire safety design, 24
remodeling buildings, 27–28
remote alarms, 29
remote area calculations, 121, 128
removed lines, symbols for, 268
repairing systems, 148
replacing
parts in systems, 148
sprinklers, 86
reports
clean-gas fire-protection system
specifications, 230–231
high-pressure CO_2 system
specifications, 219
inspections, 147
reservoirs, pump sizing calculations,
98–103
residential fire extinguishers, 170
residential sprinklers, 80, 81, 149

residual pressure
defined, 28
hydrant tests, 125
resistance coefficient (K), 97, 101,
124, 131, 261
revolutions per minute (pumps), 46,
57
Reynolds, Osborne, 92
Reynold's number for turbulence, 92
right-hand rotation pumps, 56
ring-nozzle orifices, 79
riser clamps, 141
risers
arrangements, 143–144
fire-department (Siamese) connections, 136
flushing, 72
sprinkler system design, 70–71,
296
standpipe systems, 61
water-deluge systems, 83
rises or drops, 268
rock sites, 27
rod couplings, 141
rods, 301
roof design, pressure drop and, 134
room geometry and congestion,
detectors and, 36
rotating electrical equipment fires,
163
rpm (revolutions per minute), 46, 57
rubber-coated floors, 174
rubber fires, 42
rubber mixing area fires, 300
rubbish disposal, 4, 8, 27

S
s (seconds), 303
Safe Drinking Water Act of 1974, 90
safety. *See also* hazards
education, 6–10
life safety assessments, 173
in water fire-protection system

specifications, 185–186
salt aerosol, 137–138
salt water, 89
salt water reservoirs, 137–138
sample fire-protection notes, 279–280
sample specifications
    clean-gas fire-protection system
        specifications, 227–237
    high-pressure CO_2 system
        specifications, 217–226
    portable fire extinguishers
        specifications, 213–216
    water fire-protection systems,
        175–210
saturation in fire-retardant treatment,
    21
scales for high-pressure cylinders,
    157–159
schedules
    fire drills, 3
    inspections, 146–147
    maintenance, 149
    tests, 149
sea water, 89
sealed valves, 140, 147, 149
seconds (s), 303
sectionalizing post-indicator valves,
    59
selecting fire extinguishers, 282–285
self-restoring detectors, 31, 32
sensitivity to potential fires, 85
sequence of operations in clean-gas
    fire-protection systems, 236
shielding on sprinkler pipes, 137
shipboard storage fires, 300
shipping dock fires, 139, 167
shop drawings. See also drawings
    clean-gas fire-protection system
        specifications, 230–231
    high-pressure CO_2 system
        specifications, 218–219
    symbols, 268–273
    water fire-protection system

specifications, 182–183
shut-off fans in carbon dioxide
    systems, 156
shut-off pressure, 46
shut-off valve switches in water fire-
    protection system specifica-
    tions, 203
shutdown alarm on pumps, 47
shutting water off, 85
SI units, 301
Siamese connections. See fire-
    department connections
side beam adjustable hangers, 141
side beam attachments, 141
sidewall sprinklers, 79, 81, 269
signatures, fires, 16, 173
signs
    for carbon dioxide systems, 156–
        157
    in water fire-protection system
        specifications, 197
simulating smoke situations, 17
single-stage pumps, 45
sizing
    booster pumps, 48
    carbon dioxide system pipes, 159–
        163
    extra hazard occupancy pipe
        sizing, 115
    light and ordinary occupancy pipe
        sizing, 114
    pipes, 94, 145
    pumps, 98–103, 107–108
    sprinkler systems, 68, 73, 113–116
    standpipe systems, 61–63
skylights, 138
sleeves
    installation, 209
    in water fire-protection system
        specifications, 197
slope of pipes, 145
smoke
    control, 17–18

detectors, 18, 30, 33–35, 38, 39
development ratings, 142
evacuation, 174
evacuation fans, 29
inhalation, 16
smoke-free areas, 17, 174
smoke-stop doors, 25
smoking areas, 27
snow (carbon dioxide), 152
sodium
as extinguisher, 21
carbon dioxide systems and, 153
portable fire extinguishers, 169
sodium bicarbonate, 167
software for hydraulic calculations,
117, 289–294
solder-link fusible components, 75,
76–77
solenoid valves, 270
solids in water, 90
solubility of water, 90
solvent cleaning tank fires, 300
solvent thinned coating fires, 300
spacing of detectors, 37
spare parts
in clean-gas fire-protection system
specifications, 230
data for high-pressure CO_2 system
specifications, 219
maintenance and, 148
spare pumps, 48–49
spare-sprinkler cabinet, 145
spare sprinklers, 139, 145, 211
spark-proof equipment, 23, 174
sparks, 15–16
specialty devices in sprinkler
systems, 211
specifications
carbon dioxide (CO_2) systems, 157
clean-gas fire-protection system
specifications, 227–237
high-pressure CO_2 system
specifications, 217–226

portable fire extinguishers
specifications, 213–216
sample fire-protection notes, 279–
280
water system sample specification,
175–210
speed of pumps, 46
spot concrete inserts, 141
spot detectors, 31
spray patterns for sprinklers, 78–80
spring-loaded check valves, 89
sprinkler alarm valves (AV), 270
sprinkler cabinets
spare-sprinkler cabinets, 145
water fire-protection system
specifications, 203
sprinkler guards and water shields,
210
sprinkler heads
area per, 85
deflectors, 79, 138
dimensions, 82
number of heads, 70, 126
number per branch lines, 126
orifice size, 124
overview, 75–77
sizing pipes to, 113–116
spare heads, 139, 145, 211
spray patterns, 78–80
symbols for, 268
temperature ratings, 77
testing and maintenance, 149
types, 79–80, 139–140
in water fire-protection system
specifications, 187–188
sprinkler room plans, 295, 296
sprinkler systems, 43
alarm valves (AV), 270
alarms, 68, 86–87, 270
altering sprinklers, 86
area of coverage, 72–73, 85, 125
arrangements, 143–144
components, 74–80

drainage, 73–74
effectiveness statistics, 68
failures, 87
flushing, 72
foam, 166
guards and water shields, 210
heads. *See* sprinkler heads
history, 66
hydraulic calculations, 117–134
hydropneumatic tanks, 50–51, 57
installation, 85–86
lines, 268
maintenance, 69
operations, 68–69
ordinary hazard sprinkler system
    plans, 126–134
overview, 65
pipe-schedule method calcula-
    tions, 116–117
piping for, 70–71
    protecting pipes, 137–138
    sizing pipes, 113–116
plans, 118–119
pressure, 71–72
in remodeled buildings, 28
spare parts, 139, 145, 211
specialty devices, 211
in specifications, 181
spray patterns, 78–80
sprinkler room plans, 295, 296
standards for, 135
strainers, 70, 122, 196–197
system design, 69–74
temperature ratings, 71–72, 77
temporary systems, 26
testing, 135, 140
types of sprinkler systems, 66–68
valves in, 140
water-deluge spray systems, 80,
    83–85
water supply, 70
Sprinkler Systems, Care, Mainte-
    nance (NFPA Standard no.

13A), 65, 135, 184
Sprinkler Systems, One- and Two-
    Family Dwellings (NFPA
    Standard no. 13D), 66
Sprinkler Systems, Residential
    Occupancies up to and
    including Four Stories in
    Height (NFPA Standard no.
    13R), 66
square-marked extinguishers, 171
square miles, 301
St (strainers), 70, 122, 196–197
stabilizers (fire), 16
stack effect, smoke and, 17, 18
stages of fires, 173
stagnant water in fire fighting
    systems, 90
stairwells
    extinguishing fires in, 300
    fire safety design, 24
    pressurization, 17, 18
    smoke and, 18
    in Triangle Shirtwaist building, 7
stale water in sprinkler systems, 71
standard 30° elbows, 89, 110
standard 45° elbows
    equivalent length, 110
    friction loss, 262
    symbols for, 269
    in water-distribution systems, 89
standard 90° elbows
    equivalent length, 106
    friction loss, 101, 262
    in water-distribution systems, 89
standard abbreviations
    hydraulic calculations, 122
    symbols list, 268–273
Standard International (SI) units, 301
Standard Methods of Fire Tests of
    Building Construction and
    Materials (NFPA Standard
    no. 251), 23
standard steel piping friction loss

tables, 254–260
standard tees
    equivalent length, 106
    friction loss, 263
*Standard Types of Building Con-
    struction (NFPA Standard no.
    220), 23*
standards. *See* codes and standards
standby generator room fire doors, 25
*Standpipe and Hose Systems (NFPA
    Standard no. 14),* 60, 179
standpipe systems, 60–63
    as fixed water systems, 43
    carbon dioxide systems, 155
    classification, 61
    defined, 60
    dry standpipe systems, 192–193
    fire protection and, 26
    sizing, 61–63
    in specifications, 180, 192–193
    symbols for, 268
star-marked extinguishers, 171
start points for jockey pumps and fire
    pumps, 49
start switches in clean-gas fire-
    protection system specifica-
    tions, 234
states
    state authorities, 1
    state building codes, 12
    state fire-prevention codes and
        standards, 12
static head
    calculating, 94
    defined, 89
    hydrant tests, 124
    pressure calculations and, 110
    pump sizing calculations, 99
    sprinkler pressure and, 134
    suction side of pumps, 101
statistics
    fire-caused deaths, 7
    fire department responses per day,

7
fire information, 10
smoke inhalation, 16
sprinkler system effectiveness, 68
steam turbines (pumps), 46
steel piping
    carbon dioxide, 153
    friction, 100
    friction loss tables, 254–260
    light and ordinary occupancy pipe
        sizing, 114–115
stop check valves, 264
stop points for jockey pumps, 49
storage
    alcohol storage area fires, 300
    clean-gas fire-protection system
        specifications, 231
    combustible materials, 4
    flammable liquids storage, 300
    flammable solids storage, 300
    fuel oil storage fires, 300
    fur-storage vaults, 162
    high-piled storage, 72
    high-pressure carbon dioxide, 153,
        221
    lignite storage area fires, 300
    *NFPA Standard no. 231 (Storage,
        General),* 69
    *NFPA Standard no. 231C (Rack
        Storage of Materials),* 69
    oil-storage room fire doors, 25
    paper storage, 162
    petrochemical storage area fires,
        300
    portable fire extinguishers
        specifications, 214
    product delivery, storage and
        handling
        clean-gas fire-protection
            system specifications,
            231, 232
        high-pressure CO_2 system
            specifications, 221

portable fire extinguishers
    specifications, 214
shipboard storage fires, 300
storage cylinders for clean-gas fire-
    protection system specifica-
    tions, 232
storage points for carbon dioxide
    systems, 163
storage room fire doors, 25
storage tank systems
    diesel fuel for pumps, 47
    hydropneumatic tanks, 50–51, 57
    water fire-suppression systems, 43
straight pipe flow, 96
strainers, 70, 122, 196–197
strategies for fighting fires, 41
submittals
    clean-gas fire-protection system
      specifications, 230–231
    high-pressure CO_2 system
      specifications, 218–219
    portable fire extinguishers
      specifications, 214
suction head, 46, 102
suction-life applications, 46
suction sides of pumps, 101
sudden contractions in pipes, 89, 110
sudden enlargements in pipes, 89, 95,
    110
summary sections in high-pressure
    CO_2 system specifications,
    217–221
summation of flow (Q), 122, 129
super K, 167
supervisory LEDs, 235
supply piping in water fire-protection
    system specifications, 189–
    190
supports and hangers
    types of, 141, 142
    in water fire-protection systems,
      74, 194–195
suppression agents. *See* extinguish-

ing agents
swing-check valves
    equivalent length, 106
    friction loss, 265
    symbols for, 122
    in water-distribution systems, 89
    in water fire-protection system
      specifications, 191
switches
    abort switches, 223–224, 234
    electric "temper" switches, 87
    flow switches, 273
    oil-filled circuit breakers or
      switches, 25
    pressure-activated alarm switches,
      86–87
    pressure switches, 49, 273
    shut-off valve switches, 203
    start switches, 234
    tamper switches, 149, 203, 272
    water flow switches, 271
    zone disable switches, 235
switchgear rooms
    carbon dioxide systems, 161–163
    fires, 300
symbols
    hydraulic calculations, 122
    list, 268–273
synthetic foam, 166
system checkout and testing in
    clean-gas fire-protection
    systems, 237
system operation in clean-gas fire-
    protection system specifica-
    tions, 231
system risers, 70–71

T
T (tees). *See* tees (T)
tags and identification, 140, 195–196,
    203
tamper switches
    symbols for, 272

testing and maintenance, 149
in water fire-protection system
    specifications, 203
tapered nozzle orifices, 79
tees (T)
    equivalent length, 106
    friction loss, 263
    symbols, 122
    in water-distribution systems, 89
temper switches, 87
temperature
    conversion factors, 303, 304–306
    detecting rapid rises, 32–33
    of fires, 173
    pumps and, 47
    sprinkler systems water supply,
        71–72
    temperature-sensitive semicon-
        ductor insulation, 31
    of water, 93
temperature control valves, 273
temperature elements, 273
temperature-sensitive semiconductor
    insulation, 31
temporary fire protection systems, 26
temporary water supplies, 28
testing
    clean-gas fire-protection systems,
        230, 237
    high-pressure CO_2 systems, 219,
        226
    overhead pipes, 208
    puff tests, 157
    scheduling tests, 149
    sprinkler systems, 135, 140
    system equipment, 148
    test connections, 269
    test pressures for sprinkler
        systems, 72
    underground pipes, 208
textiles. *See* cloth or textiles
thermal detectors. *See* heat detectors
thermal expansion, smoke and, 17

thermal shock, 152
three-dimensional Class B fires, 288
three-phase motors, 46
three-way valves, 270
tilted deflectors, 83
tilting disc check valves, 265
timers (pumps), 47
titanium
    carbon dioxide systems and, 153
    portable fire extinguishers, 169
torque, 301
total discharge head loss, 102–103
total flooding systems
    dry chemicals, 167
    pre-installation information lists,
        275–277
    sizing pipes, 159–160
    types of carbon dioxide systems,
        155
total friction, 97
total pressure drop, 101, 132, 133–134
total pressure (P_t), 46, 122, 130
total suction head, 102
transformers, 25, 300
transportation systems as hazards,
    174
trash disposal, 4, 8, 27
tree arrangement (sprinkler systems),
    143–144
triangle-marked extinguishers, 171
Triangle Shirtwaist fire, 7–9
triple-point pressure, 151
trouble signals in fire detection
    systems, 30
turbine lubricating oil fires, 300
turbulent flow in pipes, 92, 93, 95
Type A foam, 42
Type K copper friction loss tables,
    247–253
Type L copper friction loss tables,
    247–253
Type M copper friction loss tables,
    247–253

types of fires, 281, 299–300

U
U-hooks, 141
UL listings
    defined, 2
    portable fire extinguishers, 171
    pumps, 45
ultraviolet detectors, 35, 38, 39, 273
uncontrolled fires, 1
underground control valves, 271
underground piping, 142
    symbols, 268
    testing and flushing, 208
    underground mains, 26
Underwriters Laboratories, Inc. (UL)
    labeling, 2
    portable fire extinguishers, 171
    pumps, 45
unheated areas and sprinkler
        systems, 68
unions, 269
unit prices in water fire-protection
        system specifications, 185
units, converting, 301–306
universal top and bottom beam
        clamps, 141, 142
upright sprinklers, 79, 81, 82, 83, 268
usage guidelines for fire extinguish-
        ers, 282

V
v (velocity). *See* velocity
valuable items in fire protection
        areas, 66, 67, 153
valve actuators in clean-gas fire-
        protection system specifica-
        tions, 233
valves. *See also specific types of*
        valves
    diagrams, 183–184
    equivalent length, 105–107
    high-pressure CO_2 system

        specifications, 221–222
    inspecting, 146
    pressure drops in water systems
        and, 99
    in sprinkler systems, 140
    tags and charts, 203
    in water-distribution systems, 89
    water fire-protection system
        specifications, 183–184,
        190–192, 203
vandalism, 147
vane-type water flow alarms, 86
variable pressure alarm check valves
        in water fire-protection system
        specifications, 198
various voltage motors, 46
vegetable oil fires, 300
velocity
    calculating, 94
    conversion factors, 303
    gravity system calculations, 110
    pumps, 46
    straight pipe flow, 96
    symbols for, 122
    of water, 93
    water in pipes, 92
velocity head
    calculating, 97
    defined, 89, 92
    velocity pressure (P_v), 122
ventilation
    detectors and, 36
    equipment fire-safety design, 23
    fire suppression and, 29
    systems as hazards, 174
venting, pressure-relief formulas,
        160–161
venturi principle, 166
vertical-burning fires, 166
vertical pipes
    pressure loss, 98
    pump sizing calculations, 98–103
vertical pumps, 45

vertical shafts in building, 139
vertical sidewall sprinklers, 79
vertical-turbine pumps, 45, 46
victaulic coupling, 138
viscosity of water, 90, 91, 93
visibility in carbon dioxide systems, 156
volts (V), 301
volume conversion factors, 303
volume per unit time (flow), 303
volutes in pumps, 46

W
wafer check valves (WCV), 122
waivers from code requirements, 12
wall brackets, 141
wall indicator valves in water fire-protection system specifications, 197–198
wall plates in water fire-protection systems specifications, 209
warning lights in high-pressure CO_2 system specifications, 224
warning signs for carbon dioxide systems, 156–157
warranty
    in clean-gas fire-protection systems, 237
    in water fire-protection system specifications, 186
waste, accumulation of, 4, 8, 27
water
    as cooling agent, 43
    as extinguishing agent, 20, 42
    density, 73, 120, 125
    discharge coefficient (K), 97, 101, 124, 131, 261
    fire pumps. *See* fire pumps
    flow in pipes, 89, 92–93
    in hydraulic calculations, 125
    hydropneumatic tanks, 50–51, 57
    non-potable fire-fighting supply, 89

pressure testing and maintenance, 149
properties, 90–93
sprinkler head discharge rates, 79
water-filled pipes, 142
water-filled portable extinguishers, 170
water-flow alarms, 39, 86
wet-type sprinkler systems, 120
wet water, 20
Water-Based Fire Protection Systems (NFPA Standard no. 25), 184
water-deluge spray systems, 80, 83–85
water-filled pipes, 142
water-filled portable extinguishers, 170
water fire-protection systems. *See also* sprinkler systems; standpipe systems
    fire-suppression systems, 59–63
    hoses, 63
    specifications, 175–210
    standpipes, 60–63
water-flow alarms, 39, 68, 86, 149
water-flow switches (WFS), 271
water hammer, preventing, 94
water meters, 137, 272
water motor gongs, 86, 87, 198
water pressure testing and maintenance, 149
water storage tanks, 50–51, 57
water supply systems
    common water supplies, 136
    fire suppression, 43–44
    information in hydraulic calculations, 124
    low-pressure water sources, 48
    mains, 60
    pipe-schedule sprinkler systems, 116
    private systems, 59
    protecting, 90

sprinkler systems, 70, 74
temporary fire protection systems,
    26, 28
*Water Tanks for Private Fire
    Protection (NFPA Standard
    no. 22)*, 43
water temperature in sprinkler
    systems, 72
water-type check valves, 89
wavelength (light), 35
WCV (butterfly wafer check valves),
    122
weather shelters, 27
weekly inspections, 146
weight of CO_2 cylinders, 159
welding
    fire safety and, 26
    operations, 4
    standards, 138
    in water fire-protection systems
        specifications, 209–210
wet-chemical systems, 168, 203–204
wet-pipe sprinkler systems, 66–67, 73,
    90, 120
wet standpipes, 61
wet water, 20
wheeled fire extinguishers, 281–294
wind speed, smoke and, 17
wood
    combustibility, 19
    extinguishing fires, 42
    portable fire extinguishers, 169
work installed or provided in water
        system sample specification,
        178
work measurements conversion
        factors, 302
workmanship, 145
worksheets for hydraulic calcula-
        tions, 133
wraparound U-hooks, 141

Y
yard and sprinkler system elevation
        plans, 119
yard distribution systems, 59
yard hydrants, 59
yard piping, 142
yards, converting to SI units, 301, 302

Z
zirconium, 169
zone disable switches, 235
zone status LEDs, 235